Multiresolution Frequency Domain Technique for Electromagnetics

Synthesis Lectures on Computational Electromagnetics

Editor
Constantine A. Balanis, *Arizona State University*

Synthesis Lectures on Computational Electromagnetics will publish 50- to 100-page publications on topics that include advanced and state-of-the-art methods for modeling complex and practical electromagnetic boundary value problems. Each lecture develops, in a unified manner, the method based on Maxwell's equations along with the boundary conditions and other auxiliary relations, extends underlying concepts needed for sequential material, and progresses to more advanced techniques and modeling. Computer software, when appropriate and available, is included for computation, visualization and design. The authors selected to write the lectures are leading experts on the subject that have extensive background in the theory, numerical techniques, modeling, computations and software development.
The series is designed to:

- Develop computational methods to solve complex and practical electromagnetic boundary-value problems of the 21st century.

- Meet the demands of a new era in information delivery for engineers, scientists, technologists and engineering managers in the fields of wireless communication, radiation, propagation, communication, navigation, radar, RF systems, remote sensing, and biotechnology who require a better understanding and application of the analytical, numerical and computational methods for electromagnetics.

Multiresolution Frequency Domain Technique for Electromagnetics
Mesut Gokten, Atef Elsherbeni, and Ercument Arvas
2012

Scattering Analysis of Periodic Structures Using Finite-Difference Time-Domain Method
Khaled ElMahgoub, Fan Yang, and Atef Elsherbeni
2012

Introduction to the Finite-Difference Time-Domain (FDTD) Method for Electromagnetics
Stephen D. Gedney
2011

Analysis and Design of Substrate Integrated Waveguide Using Efficient 2D Hybrid Method
Xuan Hui Wu and Ahmed A. Kishk
2010

An Introduction to the Locally-Corrected Nyström Method
Andrew F. Peterson and Malcolm M. Bibby
2009

Transient Signals on Transmission Lines: An Introduction to Non-Ideal Effects and Signal
Integrity Issues in Electrical Systems
Andrew F. Peterson and Gregory D. Durgin
2008

Reduction of a Ship's Magnetic Field Signatures
John J. Holmes
2008

Integral Equation Methods for Electromagnetic and Elastic Waves
Weng Cho Chew, Mei Song Tong, and Bin Hu
2008

Modern EMC Analysis Techniques Volume II: Models and Applications
Nikolaos V. Kantartzis and Theodoros D. Tsiboukis
2008

Modern EMC Analysis Techniques Volume I: Time-Domain Computational Schemes
Nikolaos V. Kantartzis and Theodoros D. Tsiboukis
2008

Particle Swarm Optimization: A Physics-Based Approach
Said M. Mikki and Ahmed A. Kishk
2008

Three-Dimensional Integration and Modeling: A Revolution in RF and Wireless Packaging
Jong-Hoon Lee and Manos M. Tentzeris
2007

Electromagnetic Scattering Using the Iterative Multiregion Technique
Mohamed H. Al Sharkawy, Veysel Demir, and Atef Z. Elsherbeni
2007

Computational Electronics
Dragica Vasileska and Stephen M. Goodnick
2006

Higher Order FDTD Schemes for Waveguide and Antenna Structures
Nikolaos V. Kantartzis and Theodoros D. Tsiboukis
2006

Introduction to the Finite Element Method in Electromagnetics
Anastasis C. Polycarpou
2006

MRTD(Multi Resolution Time Domain) Method in Electromagnetics
Nathan Bushyager and Manos M. Tentzeris
2006

Mapped Vector Basis Functions for Electromagnetic Integral Equations
Andrew F. Peterson
2006

Multiresolution Frequency Domain Technique for Electromagnetics

Mesut Gokten, Atef Elsherbeni, and Ercument Arvas

ISBN: 978-3-031-00586-2   paperback
ISBN: 978-3-031-01714-8   ebook

DOI 10.1007/978-3-031-01714-8

A Publication in the Springer series
SYNTHESIS LECTURES ON COMPUTATIONAL ELECTROMAGNETICS

Lecture #29
Series Editor: Constantine A. Balanis, *Arizona State University*
Series ISSN
Synthesis Lectures on Computational Electromagnetics
Print 1932-1252   Electronic 1932-1716

Multiresolution Frequency Domain Technique for Electromagnetics

Mesut Gokten
Turksat AS R&D and Satellite Design Department

Atef Elsherbeni
The University of Mississippi

Ercument Arvas
Syracuse University

SYNTHESIS LECTURES ON COMPUTATIONAL ELECTROMAGNETICS #29

ABSTRACT

In this book, a general frequency domain numerical method similar to the finite difference frequency domain (FDFD) technique is presented. The proposed method, called the multiresolution frequency domain (MRFD) technique, is based on orthogonal Battle-Lemarie and biorthogonal Cohen-Daubechies-Feauveau (CDF) wavelets. The objective of developing this new technique is to achieve a frequency domain scheme which exhibits improved computational efficiency figures compared to the traditional FDFD method: reduced memory and simulation time requirements while retaining numerical accuracy.

The newly introduced MRFD scheme is successfully applied to the analysis of a number of electromagnetic problems, such as computation of resonance frequencies of one and three dimensional resonators, analysis of propagation characteristics of general guided wave structures, and electromagnetic scattering from two dimensional dielectric objects. The efficiency characteristics of MRFD techniques based on different wavelets are compared to each other and that of the FDFD method. Results indicate that the MRFD techniques provide substantial savings in terms of execution time and memory requirements, compared to the traditional FDFD method.

KEYWORDS

multiresolution analysis, wavelets, MRFD, finite difference technique, frequency domain, computational electromagnetics

Contents

CHAPTER 1

Introduction

In the opening chapter, the background and motivation of this work is explained. The organization of the book is also provided.

1.1 BACKGROUND AND MOTIVATION

Although there seem to be many weapons in an engineer's or scientist's arsenal to model and solve real life problems, these can be categorized into just three classes; namely experimental, analytical, or numerical techniques. Historically, analytical and experimental techniques have been the primary tools of scientific progress, but these techniques have their limits. Experiments are usually very expensive, time consuming, and sometimes hazardous. The range of problems that can be solved analytically is surprisingly limited due to the complex geometries defining practical problems. Fortunately, numerical techniques usually deliver when other techniques fail.

After the invention of digital computers, interest in numerical solutions to electromagnetic problems boomed in the 1960s. Since then, very complex problems have been solved with these methods, most of which would be impossible to solve by closed form analytical techniques. The numerical approach is not only able to solve otherwise impossible problems, it also has the advantage of being low cost and fast.

The performance of modern computers is nowhere close to the early examples of the 1960s. Advances in semiconductor technology resulted in much faster CPUs and very large data storage capabilities, which made it feasible to characterize large and sophisticated microwave problems. However, development of even complex circuits such as monolithic microwave integrated circuits (MMIC) or multilayer low temperature cofired ceramic (LTCC) circuits increased the burden on the computational techniques despite the advances in computer technology. It is thus necessary to improve the efficiency of the currently available computational techniques or develop completely new techniques which make it possible to utilize computer resources more efficiently.

Computational electromagnetic techniques model microwave circuits either in time domain or frequency domain. Frequency domain techniques are preferred when information over a narrow frequency range is pursued. These techniques also became suitable for broadband applications with the introduction of frequency sweep techniques such as asymptotic waveform equation [1]. This book focuses on the solution to Maxwell's equations in frequency domain by developing new schemes which resemble the finite difference frequency domain (FDFD) method.

The finite difference frequency domain technique provides a mathematically straightforward analysis method to characterize arbitrary geometries with different material types. The advantages of this technique are its simplicity, stability, efficiency, and ease of implementation. However, despite its advantages, this technique suffers from some limitations while treating electrically large problems or fine detailed structures, due to the substantial computational resources required. The limitations are mostly the result of the uniform spacing of grid points in the lattice and second order accuracy of the central difference approximation. Multiresolution analysis techniques have the potential to address both problems by offering much more accurate higher order schemes equipped with the capability of non-uniform gridding.

Over the last decade, multiresolution analysis techniques have successfully been applied to various computational electromagnetic methods yielding significant computational CPU time and memory savings, compared to the traditional techniques. Multiresolution analysis has found application in the improvement of method of moments (MoM) [2, 3, 4] by generating a sparse linear system. The transmission line matrix (TLM) method [5] is also improved with multiresolution analysis techniques.

The finite difference time domain (FDTD) method benefited most from multiresolution techniques and evolved into the multiresolution time domain (MRTD) technique. Various MRTD schemes [6, 7, 8, 9] based on a number of different wavelets have been developed which have better dispersion characteristics compared to the FDTD technique.

The FDFD method, on the other hand, has not yet benefited from the advantages of multiresolution analysis. In this work, we formulate general 3D frequency domain numerical methods based on orthogonal and biorthogonal multiresolution analysis from which a special case leads to the FDFD method.

1.2   BOOK OVERVIEW

The organization of this book is as follows:

In Chapter 2, brief overviews of finite difference schemes and multiresolution analysis are presented, so that the reader will be better prepared to apprehend the subjects presented in subsequent chapters.

Chapter 3 presents the derivation of biorthogonal wavelet based multiresolution frequency domain schemes. Also included are the derivation of the finite difference frequency domain method from MoM formulation and the selection criteria of the appropriate wavelet bases.

The application of the newly developed MRFD schemes to closed space problems is performed in Chapter 4. In order to verify and to demonstrate the efficiency of the new MRFD schemes, one, two and three-dimensional closed space problems are considered in this chapter. The structures that are analyzed in this chapter are all assumed to be enclosed by a perfect electric conductor (PEC) boundary; hence the name closed space structures. For the purpose of supporting the claim that MRFD formulation is inherently more efficient than the FDFD scheme, the closed space problems are solved with the aid of both methods and the numerical results are compared. The treatment of PEC boundary conditions in the context of the image principle and the multiple image technique is also addressed in this chapter.

The MRFD technique is used to characterize open space problems in Chapter 5. A scattered field formulation is developed to model open problems. The perfectly matched layer (PML) technique is employed in order to terminate the computational space. Numerical results of 2D scattering and radiation problems are presented at the end of the chapter.

In Chapter 6, a new MRFD scheme, called the inhomogeneous MRFD, for analyzing inhomogeneous problems, is formulated. A one-dimensional closed space problem is considered in order to evaluate the new formulation.

The book is concluded in Chapter 7. The contributions of this work are summarized and some recommendations for future work are provided.

BIBLIOGRAPHY

CHAPTER 2

Basics of the Finite Difference Method and Multiresolution Analysis

In this chapter, brief overviews of finite difference schemes and multiresolution analysis are presented, so that the reader will be better prepared to apprehend the subjects presented in subsequent chapters.

2.1 OVERVIEW OF THE FINITE DIFFERENCE METHOD

In general, Maxwell's equations are used to solve electromagnetic problems. However, only few problems with simple geometries can be solved satisfactorily by classic analytic techniques. Real life electromagnetic problems are usually made of complex structures, the materials that are utilized may be anisotropic and the boundary conditions may be mixed. In such instances, closed form analytic approaches fail to generate successful solutions. Consequently, one must resort to numerical methods whenever a problem with such complexity arises.

Numerous numerical methods are available for solving partial differential equations (PDEs), some of which are the method of moments (MoM) [10], the finite element method (FEM) [11], the transmission line method (TLM) [12], finite difference time domain method (FDTD) [13], and the finite difference frequency domain method (FDFD) [14, 15].

Every one of these methods has its own unique strengths and weaknesses, depending on the problem at hand. One of the strengths of the FDFD scheme is that it has no analytical load, such as calculation of structure-dependent Green's functions, and thus it is easy to understand and implement the method. It is easy to model complex materials such as anisotropic materials or frequency dependent materials, so material generality of the FDFD technique is considered to be good. The technique is also very robust and does not suffer from stability problems often encountered in time domain methods such as FDTD or TLM. A qualitative comparison of the FDFD method with various numerical methods can be found in Table 2.1 [16].

The finite difference method was first introduced to model nonlinear hydrodynamic equations [17] in the 1960s and was named "the method of squares." Initial work of Yee [18] brought the method into the field of electromagnetic boundary value problems. Yee's method discretized Maxwell's equations such that the values of electric and magnetic fields are sampled at suitable positions

Table 2.1: Comparison of various computational numerical techniques

Numerical Technique	Finite Difference	Finite Element	Method of Moments	Boundary Element
Analytical Load	Nil	Small	Large	Moderate
Require Absorbing Boundaries	Yes	Yes	No	No
Material Generality	Good	Very good	Limited	Limited
Arbitrary Contours	Poor	Good	Poor	Good
Geometric Versatility	Good	Very good	Good	Good
Type of System Equations	Sparse	Sparse	Dense	Dense
Memory Requirements	Large	Large	Moderate	Moderate
Simulation Time	Large	Large	Moderate	Moderate

of time and space in order to simulate the procedure of electromagnetic propagation. Absorbing boundaries had not yet been introduced, so Yee made his computational region finite by computing the scattering from a conducting post in an ideal conducting cavity and used an impulse function as the incident wave.

Since then, numerous researchers made significant contributions and the method has found applications in solving many different field problems. Some of these applications are the use of the finite-difference Green's function method for solving time-harmonic waveguide scattering problems involving metallic obstacles [19], the numerical calculations of absorbed energy deposition for a block model of man [20], the analysis of three-dimensional finite difference frequency domain scattering computations [14, 21], and scattering from chiral materials [22].

The finite difference techniques are based on approximating the spatial and temporal derivatives by finite difference equations. Finite difference approximations are algebraic in form. They relate the value of the dependent variable at a point in the solution region to the values at some neighboring points. Thus, a finite difference solution basically involves three steps [23]:

1. Dividing the solution region into a grid of nodes.

2. Approximating the given differential equation by the finite difference equivalent that relates the dependent variable at a point in the solution to its values at the neighboring points.

3. Solving the difference equations subject to the prescribed boundary conditions and/or initial conditions.

The course of action taken in the three steps is dictated by the nature of the problem being solved, the solution region, and the boundary conditions. Once the fields within the solution region are determined, additional steps may be involved such as the calculation of far field radiation or scattering parameters.

THE FINITE DIFFERENCE APPROXIMATION

The finite difference solution procedure of PDEs is based on the construction of a finite difference approximation for the derivative of a function. This approximation essentially involves estimating derivatives numerically.

The derivative of a function $f(x)$ at point x_0 can be approximated in different ways. Common methods of constructing the finite difference approximation are illustrated by the aid of Figure 2.1. Utilizing the slope of arc BC gives the forward-difference formula:

$$\frac{df(x_0)}{dx} = f'(x_0) \approx \frac{f(x_0 + \Delta x) - f(x_0)}{\Delta x}.$$ (2.1)

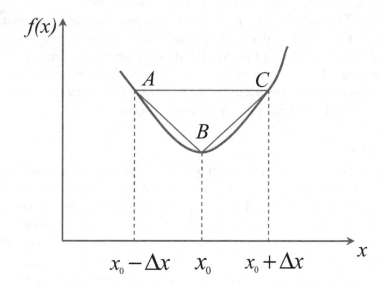

Figure 2.1: Finite difference discretization of function $f(x)$.

Estimation via the slope of arc AB results in the backward-difference formula:

$$\frac{df(x_0)}{dx} = f'(x_0) \approx \frac{f(x_0) - f(x_0 - \Delta x)}{\Delta x}$$ (2.2)

and employing the slope of arc AC yields the central-difference formula:

$$\frac{df(x_0)}{dx} = f'(x_0) \approx \frac{f(x_0 + \Delta x) - f(x_0 - \Delta x)}{2\Delta x}. \tag{2.3}$$

The above explanation of the concept of the finite difference approximation is performed in a visually intuitive manner. A mathematically correct approach can be developed by employing Taylor series expansion. Consider the Taylor expansions of $f(x_0 + \Delta x)$ and $f(x_0 - \Delta x)$:

$$f(x_0 + \Delta x) = f(x_0) + \Delta x f'(x_0) + \frac{(\Delta x)^2 f''(x_0)}{2!} + \frac{(\Delta x)^3 f'''(x_0)}{3!} + \cdots \tag{2.4}$$

$$f(x_0 - \Delta x) = f(x_0) - \Delta x f'(x_0) + \frac{(\Delta x)^2 f''(x_0)}{2!} - \frac{(\Delta x)^3 f'''(x_0)}{3!} + \cdots \tag{2.5}$$

Subtracting (2.5) from (2.4) and dividing by $2\Delta x$ yields:

$$f'(x_0) = \frac{f(x_0 + \Delta x) - f(x_0 - \Delta x)}{2\Delta x} + O[(\Delta x)^2]. \tag{2.6}$$

The first term on the right-hand side of (2.6) is the central-difference approximation of $f'(x)$ at point x_0. The second term represents the error between the approximation and the exact value of the derivative. It is clear that the error introduced by this estimation is proportional to the square of the finite difference cell size Δx; therefore the central-difference scheme is considered second order accurate. Using a similar procedure, it can be shown that the forward-difference and backward-difference schemes are first order accurate. Due to its better accuracy, central-difference approximation is usually preferred in finite difference methods.

THE YEE CELL

The first step in the construction of the finite difference algorithm is the discretization of the computational space into unit cells and the definition of the locations of the electric and magnetic field vectors associated with each cell. Yee [18] developed an algorithm in which the electric and magnetic field vector components are located in a staggered fashion as shown in Figure 2.2. The reason for the staggered grid is that when the curl operator is approximated using a difference formula, the resulting derivative is evaluated at a point that is in between the sample locations used in the difference formula. In each cell, three electric field components and three magnetic field components are defined. They do not coincide with the nodes *(i, j, k)* of the Cartesian grid. The electric field components are located at the centers of the edges of each cell, and the magnetic field components are normal to the centers of the faces. This special configuration depicts Faraday's Law and Ampere's Law. In Figure 2.2, it can be seen that each magnetic field vector component is surrounded by four electric field components forming a loop around it and simulating Faraday's Law, and each electric field vector component is surrounded by four magnetic field components forming a loop around

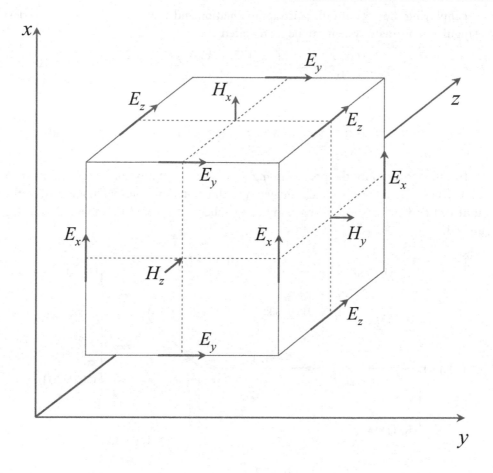

Figure 2.2: Positions of the field components on a unit cell of Yee lattice.

it and simulating Ampere's Law. Using this scheme, one can describe the explicit finite difference approximation of Maxwell's equations.

Following Yee's notation for the finite difference procedure, one may denote a space point in a rectangular lattice as:

$$(i, j, k) \triangleq (i\Delta x, j\Delta y, k\Delta z). \tag{2.7}$$

Here, Δx, Δy, and Δz are the lattice spatial increment steps along x, y, and z-axes, respectively, and i, j, and k are integers. Any function of space evaluated at a lattice point can then be represented as:

$$f(i, j, k) = f(i\Delta x, j\Delta y, k\Delta z). \tag{2.8}$$

Employing the central-difference approximation and Yee's notation, the spatial derivatives in a rectangular coordinate system can be represented by:

$$\frac{\partial f(i, j, k)}{\partial x} = \frac{f(i + 1/2, j, k) - f(i - 1/2, j, k)}{\Delta x} \tag{2.9a}$$

$$\frac{\partial f(i, j, k)}{\partial y} = \frac{f(i, j + 1/2, k) - f(i, j - 1/2, k)}{\Delta y} \tag{2.9b}$$

$$\frac{\partial f(i, j, k)}{\partial z} = \frac{f(i, j, k + 1/2) - f(i, j, k - 1/2)}{\Delta z} . \tag{2.9c}$$

In order to simplify the programming process, the notation of Yee is modified so that half integral values of i, j, and k are eliminated. For this notation, the spatial location of the fields on the unit cell and the corresponding numbering scheme are listed in Table 2.2 and illustrated in Figure 2.3.

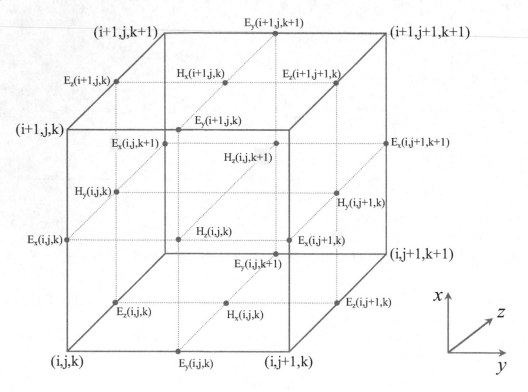

Figure 2.3: Spatial locations and numbering scheme of the field components.

The positions of the material parameters (ε and μ) are associated with field components: values of ε are associated with $\overrightarrow{E}$ field components; values of μ are associated with $\overrightarrow{H}$ field components.

Table 2.2: Actual spatial location of the indexed material and field components

Spatial Location	Corresponding Index
$E_x(i+\frac{1}{2},j,k),\ \varepsilon_x(i+\frac{1}{2},j,k)$	$E_x(i,j,k),\ \varepsilon_x(i,j,k)$
$E_y(i,j+\frac{1}{2},k),\ \varepsilon_y(i,j+\frac{1}{2},k)$	$E_y(i,j,k),\ \varepsilon_y(i,j,k)$
$E_z(i,j,k+\frac{1}{2}),\ \varepsilon_z(i,j,k+\frac{1}{2})$	$E_z(i,j,k),\ \varepsilon_z(i,j,k)$
$H_x(i,j+\frac{1}{2},k+\frac{1}{2}),\ \mu_x(i,j+\frac{1}{2},k+\frac{1}{2})$	$H_x(i,j,k),\ \mu_x(i,j,k)$
$H_y(i+\frac{1}{2},j,k+\frac{1}{2}),\ \mu_y(i+\frac{1}{2},j,k+\frac{1}{2})$	$H_y(i,j,k),\ \mu_y(i,j,k)$
$H_z(i+\frac{1}{2},j+\frac{1}{2},k),\ \mu_z(i+\frac{1}{2},j+\frac{1}{2},k)$	$H_z(i,j,k),\ \mu_z(i,j,k)$

The spatial location of the material parameters on the unit cell and the corresponding numbering scheme are also listed in Table 2.2.

The field placement and numbering schemes used in the MRFD technique are the same without any changes.

In Cartesian coordinates, Maxwell's time-harmonic curl equations lead to the following six scalar equations:

$$j\omega\varepsilon_x(x,y,z)E_x(x,y,z) = \frac{\partial H_z(x,y,z)}{\partial y} - \frac{\partial H_y(x,y,z)}{\partial z} \tag{2.10a}$$

$$j\omega\varepsilon_y(x,y,z)E_y(x,y,z) = \frac{\partial H_x(x,y,z)}{\partial z} - \frac{\partial H_z(x,y,z)}{\partial x} \tag{2.10b}$$

$$j\omega\varepsilon_z(x,y,z)E_z(x,y,z) = \frac{\partial H_y(x,y,z)}{\partial x} - \frac{\partial H_x(x,y,z)}{\partial y} \tag{2.10c}$$

$$j\omega\mu_x(x,y,z)H_x(x,y,z) = -\frac{\partial E_z(x,y,z)}{\partial y} + \frac{\partial E_y(x,y,z)}{\partial z} \tag{2.10d}$$

$$j\omega\mu_y(x,y,z)H_y(x,y,z) = -\frac{\partial E_x(x,y,z)}{\partial z} + \frac{\partial E_z(x,y,z)}{\partial x} \tag{2.10e}$$

$$j\omega\mu_z(x,y,z)H_z(x,y,z) = -\frac{\partial E_y(x,y,z)}{\partial x} + \frac{\partial E_x(x,y,z)}{\partial y}. \tag{2.10f}$$

Using Yee's lattice and the central-difference approximation, FDFD scheme can be applied to these equations as:

$$j\omega\varepsilon_x(i,j,k)E_x(i,j,k) = \frac{H_z(i,j,k) - H_z(i,j-1,k)}{\Delta y} - \frac{H_y(i,j,k) - H_y(i,j,k-1)}{\Delta z}$$

(2.11a)

$$j\omega\varepsilon_y(i,j,k)E_y(i,j,k) = \frac{H_x(i,j,k) - H_x(i,j,k-1)}{\Delta z} - \frac{H_z(i,j,k) - H_z(i-1,j,k)}{\Delta x}$$

(2.11b)

$$j\omega\varepsilon_z(i,j,k)E_z(i,j,k) = \frac{H_y(i,j,k) - H_y(i-1,j,k)}{\Delta x} - \frac{H_x(i,j,k) - H_x(i,j-1,k)}{\Delta y}$$

(2.11c)

$$j\omega\mu_x(i,j,k)H_x(i,j,k) = \frac{E_y(i,j,k+1) - E_y(i,j,k)}{\Delta z} - \frac{E_z(i,j+1,k) - E_z(i,j,k)}{\Delta y}$$

(2.11d)

$$j\omega\mu_y(i,j,k)H_y(i,j,k) = \frac{E_z(i+1,j,k) - E_z(i,j,k)}{\Delta x} - \frac{E_x(i,j,k+1) - E_x(i,j,k)}{\Delta z}$$

(2.11e)

$$j\omega\mu_z(i,j,k)H_z(i,j,k) = \frac{E_x(i,j+1,k) - E_x(i,j,k)}{\Delta y} - \frac{E_y(i+1,j,k) - E_y(i,j,k)}{\Delta x}.$$

(2.11f)

2.2   OVERVIEW OF MULTIRESOLUTION ANALYSIS

Wavelets are brief oscillating waveforms that usually last a few cycles at most. They provide convenient sets of basis functions for function spaces. The word wavelet originated from the French term *ondelette* which literally means "small waves." Smallness refers to the fact that they are localized in time, in contrast to Fourier basis functions which are perfectly localized in frequency domain but are infinite at time domain. Wavelets decay to zero as $t \to \pm\infty$ and enjoy good localization properties in frequency space. This leads to the possibility that wavelets are better suited to represent functions that are localized both in time and frequency. In particular, wavelets can represent functions with sharp spikes or edges with fewer terms. This efficiency in representing functions has obvious advantages in many applications; for example, in numerical analysis where this sparseness is a huge benefit in terms of computational speed for certain classes of problems.

The wavelet theory provides the wavelet expansion which can be used to approximate any function in the Hilbert space of square integrable functions, $L^2(R)$, in the form of:

$$f(t) = \sum_k c_k \phi(t-k) + \sum_j \sum_k d_{j,k} \psi(2^j t - k) \,.$$

(2.12)

In (2.12), the first sum represents the projection of $f(t)$ onto a subspace V_0 that corresponds to an approximation of f at a coarse level of resolution. This subspace is called the approximation space. It is generated by orthogonal translations of $\phi(t)$, which is called the scaling function.

The resolution of V_0 is successively refined by the second sum, which consists of projections of f onto the subspaces W_j, each one being spanned by a wavelet basis $\{\psi(2^j t - k)\}$. The function $\psi(t)$ is called the wavelet function or mother wavelet and the subspace W_j is called the detail space.

The wavelet expansion shows that a function can be viewed as the combination of a coarse background function and some fine details on top of it. The distinction between the coarse part and the details is determined by the resolution. At a given resolution, a signal is approximated by ignoring all details below that scale. It is possible to increase or reduce the resolution by adding or removing finer details. The word "multiresolution" refers to the simultaneous presence of different resolutions.

Figure 2.4 shows the successive approximation of an arbitrary function by step functions. At the first step, the function is approximated by the approximation space V_0. Since resolution is very rough (i.e., the step function is wide), a narrow and sharp spike at the center is skipped and not included in the approximation. At the second step, components from the detail subspace W_0 are added to the approximation. The sharp spike which was missed by the first approximation is now added to the projection. Note that the resolution is increased only where it is needed, thus the needed information is added to the approximation without adding too many components to the expansion. Two resolutions exist simultaneously in the final approximation, and thus using the term multiresolution is appropriate.

DEFINITION OF MULTIRESOLUTION ANALYSIS

The space of square integrable functions is considered. The subspace of functions that contain signal information down to scale $1/2^j$ is donated by V_j. The decomposition of the function space into a collection of subspaces V_j (denoted by $\{V_j, j \in Z\}$) is called a multiresolution if $\{V_j, j \in Z\}$ satisfies certain conditions.

The first condition is that the subspace V_j be contained in all the higher subspaces. If we denote the approximation of $f(t)$ at level j by $f_j(t)$, then $f_j(t) \in V_j$. Since information at resolution level j is necessarily included in the information at a higher resolution, V_j must be contained in V_{j+1}, that is, mathematically, $V_j \subset V_{j+1}$ for all j. The difference between $f_{j+1}(t)$ and $f_j(t)$ is the additional information about details at scale $1/2^{j+1}$ which is denoted by $d_j(t)$, so $f_{j+1}(t) = f_j(t) + d_j(t)$.

Subspaces can be decomposed accordingly:

$$V_{j+1} = V_j \oplus W_j . \tag{2.13}$$

In (2.13), the detail space W_j is the orthogonal complement of V_j in V_{j+1} which means that the inner product between any element in W_j and any element in V_j vanishes. Clearly, both V_j and W_j are subspaces of V_{j+1}, and V_{j+1} corresponds to a level of resolution twice that of V_j.

Recursively,

$$V_{j+1} = V_j \oplus W_j = V_{j-1} \oplus W_{j-1} \oplus W_j = \cdots\cdots = V_0 \oplus W_0 \oplus W_1 \oplus \cdots \oplus W_j . \tag{2.14}$$

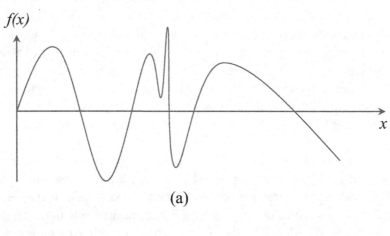

(a)

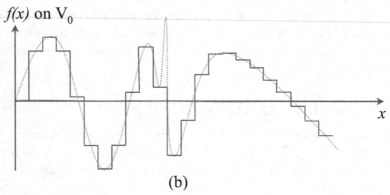

(b)

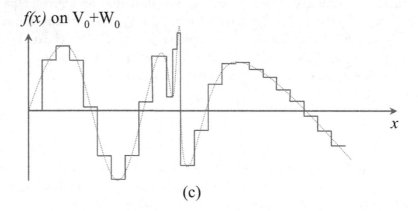

(c)

Figure 2.4: Projections of a function $f(x)$ onto subspaces V_0 and $V_0 + W_0$.

It is also worthwhile to mention that two detail spaces at different resolutions are orthogonal, and the detail space W_j is orthogonal to an approximation space $V_{j'}$ only when the detail space has higher resolution.

The second condition for multiresolution analysis is that at the finest resolution, all square integrable functions are included and at the coarsest resolution, only the zero function exists. When the resolution is increased, more and more details are included in the approximation and as the resolution goes to infinity, the entire initial space $L^2(R)$ should be recovered, that is, $\lim_{j\to\infty} V_j = L^2R$ or $\overline{UV_j} = L^2(R)$.

As the resolution gets coarser, more details are removed, and at the coarsest resolution, $j \to -\infty$, only a constant function can survive and since it has to be square integrable, it can only be the zero function. Therefore, as a third condition, $\lim_{j\to-\infty} V_j = \{0\}$ or $\cap V_j = \{0\}$.

The fourth condition is the scaling condition which states that a function $f(x)$ belongs to V_j if and only if $f(2x)$ belongs to V_{j+1}.

The fifth condition is the shift invariance of the space V_j. If $f(t)$ is a function defined in V_j, its translates by integers, $f(t-k)$, can also be defined in V_j.

The final requirement is the function ϕ belongs to V_0 and the set $\{\phi(t-k), k \in Z\}$ is an orthonormal basis (using the L^2 inner product) for V_0. Similarly, if we define $\phi_{jk}(t) = \phi(2^j t - k)$, then $\{\phi_{jk}(t)\}$ forms an orthonormal basis for V_j.

In summary, a multiresolution analysis of $L^2(R)$ is a nested sequence of subspaces V_j such that:

1. $... \subset V_{-1} \subset V_0 \subset V_1 \subset V_2 ... \subset L^2(R)$ (nested)

2. $\cap V_j = \{0\}$ (separation)

3. $\overline{UV_j} = \{0\}$ (density)

4. $f(t) \in V_j \Leftrightarrow f(2t) \in V_{j+1}$ (scaling)

5. $f(t) \in V_j \Rightarrow f(t-k) \in V_j$ (shifting)

6. There exists a function $\phi(t)$, called the scaling function, such that $\{\phi(t-k)\}$ forms an orthonormal basis of V_0.

$V_0 \subset V_1$, so any function in V_0 can be expanded in terms of the basis functions of V_1. Since the scaling function is in V_0 it also can be expanded in terms of $\{\phi_{1k}(t)\}$:

$$\phi(t) = \sum_k h_k \phi_{1k}(t) = \sum_k h_k \phi(2t - k) . \tag{2.15}$$

This equation is called the scaling relation or dilation relation. It relates the scaling function at two successive scales, and, hence, it is also referred to as the two-scale equation or refinement equation. $\{\phi_{1k}(t)\}$ are orthonormal, thus, the coefficients h_k can be calculated by the aid of inner product:

$$h_k = \langle \phi_{1k}, \phi \rangle = \int_{-\infty}^{\infty} \phi(t)\phi(2t - k)dt \ . \tag{2.16}$$

Some properties of h_k are listed below:

$$\sum_{k \in Z} |h_k|^2 = 2 \tag{2.17a}$$

$$\sum_{k \in Z} h_k = 2 \tag{2.17b}$$

$$\sum_{k \in Z} h_{2k} = 1 \ \text{ and } \ \sum_{k \in Z} h_{2k+1} = 1 \tag{2.17c}$$

$$\sum_{k \in Z} h_{k+2l} h_k = 2\delta_{0,l} \ . \tag{2.17d}$$

Since $\{\psi(t - k)\}$ is in W_0 and $W_0 \subset V_1$, mother wavelet can be written as a superposition of the basis functions for V_1:

$$\psi(t) = \sum_k g_k \phi_{1k}(t) = \sum_k g_k \phi(2t - k) \ . \tag{2.18}$$

This equation is called the wavelet relation. It relates the mother wavelet to the scaling function at the next finer scale. Again, using the orthonormality of $\{\phi_{1k}(t)\}$, the coefficients g_k can be computed by the aid of the inner product:

$$g_k = \langle \phi_{1k}, \psi \rangle = \int_{-\infty}^{\infty} \psi(t)\phi(2t - k)dt \ . \tag{2.19}$$

CONSTRUCTION OF WAVELETS

It is possible to construct wavelets according to certain specifications. As a simple example, the construction of a smooth wavelet with compact support is presented. In order for the wavelet to have smoothness and compact support, the corresponding mother wavelet needs to have vanishing moments:

$$\int_{-\infty}^{\infty} \psi(t)t^p dt = 0, \qquad p = 0, 1 \ . \tag{2.20}$$

This implies that:

$$\sum_{k \in Z} (-1)^k h_{L-k-1} k^p = 0, \quad p = 0, 1 \ . \tag{2.21}$$

Here, L is the number of filter coefficients. For a four-coefficient scaling function ($L = 4$), the conditions imposed by (2.17a), (2.17b), and (2.21) translate into four equations:

$$h_0 + h_1 + h_2 + h_3 = 2 \qquad (2.22a)$$

$$h_0^2 + h_1^2 + h_2^2 + h_3^2 = 2 \qquad (2.22b)$$

$$-h_0 + h_1 - h_2 + h_3 = 0 \qquad (2.22c)$$

$$-3h_0 + 2h_1 - h_2 = 0 \qquad (2.22d)$$

leading to:

$$
\begin{aligned}
h_0 &= \tfrac{1+\sqrt{3}}{4}, & h_1 &= \tfrac{3+\sqrt{3}}{4}, \\
h_2 &= \tfrac{3-\sqrt{3}}{4}, & h_3 &= \tfrac{1-\sqrt{3}}{4}.
\end{aligned}
\qquad (2.23)
$$

Then, using the dilation relation, the scaling function becomes:

$$\phi(t) = h_0\phi(2t) + h_1\phi(2t - 1) + h_2\phi(2t - 2) + h_3\phi(2t - 3). \qquad (2.24)$$

In general, it is not possible to solve this equation directly to find the scaling function $\phi(t)$. An iterative solution is possible when $\phi(t)$ is approximated as

$$\phi_j(t) = h_0\phi_{j-1}(2t) + h_1\phi_{j-1}(2t - 1) + h_2\phi_{j-1}(2t - 2) + h_3\phi_{j-1}(2t - 3), \qquad (2.25)$$

and the iteration is started with a pulse function ($\phi_0(t) = 1$ for $0 < t < 1$). The iteration is executed until $\phi_j(t)$ is indistinguishable from $\phi_{j-1}(t)$. It is clear that the resulting wavelet does not have a closed form expression. The wavelet constructed above is called the Daubechies D4 wavelet, as the method and wavelet is developed by Daubechies. In practice however, the wavelet construction conveniently takes place in Fourier domain and the subject is rather complex to mention in a short introduction. The reader is referred to [24] for more details.

BIORTHOGONAL WAVELETS

There exists a large selection of wavelet families depending on the choice of the mother wavelet. However, desirable properties such as symmetry, compactness of support, rapid decay, and smoothness impose a variety of restrictions. The selection criteria of the optimal choice of the wavelet basis depend on the application at hand and will be addressed later in the following chapter. A biorthogonal wavelet basis is mainly used in this work, so the subject of biorthogonal wavelets is also briefly explained here.

A basis that spans a space does not have to be orthogonal. In order to gain more degrees of freedom in the construction of wavelet bases, the orthogonality condition is relaxed, allowing the development of nonorthogonal wavelet bases. One additional degree of freedom introduced by nonorthogonal wavelets is the possibility to construct symmetric, compact, and smooth wavelet functions.

The introduction of a dual basis and a dual space makes it possible to decompose a function as a linear combination or superposition of nonorthogonal basis functions. For a nonorthogonal basis $\{\phi_i(t)\}$ of a function space, one can introduce a dual basis $\{\tilde{\phi}_i(t)\}$ such that these two are orthogonal to each other:

$$\left\langle \tilde{\phi}_i(t)\phi_j(t) \right\rangle = \int_{-\infty}^{\infty} \tilde{\phi}_i(t)\phi_j(t)dt = \delta_{i,j} . \tag{2.26}$$

A function $f(t)$ can be decomposed as a superposition of the nonorthogonal basis $\{\phi(t)\}$:

$$f(t) = \sum_k c_k \phi_k(t) . \tag{2.27}$$

Here, the coefficients c_k can be calculated by the inner product incorporating the dual scaling function:

$$c_k = \left\langle \tilde{\phi}_k(t), f(t) \right\rangle . \tag{2.28}$$

It is assumed that the function space and its dual are the same, a condition satisfied in L^2. Therefore, the roles of dual basis and the original basis can be interchanged:

$$f(t) = \sum_k \langle \phi_k(t), f(t) \rangle \, \tilde{\phi}_k(t) . \tag{2.29}$$

The dilations and translations of the scaling functions $\{\phi_{jk}(t)\}$ and $\{\tilde{\phi}_{jk}(t)\}$ constitute a basis for dual approximation spaces V_j and $\tilde{V}_j$, respectively. Similarly, one can generate dual detail spaces W_j and $\tilde{W}_j$ from the dilations and translations of dual mother wavelets, $\{\psi_{jk}(t)\}$ and $\{\tilde{\psi}_{jk}(t)\}$. With the dual scaling and wavelet functions, a biorthogonal multiresolution analysis can be defined, which satisfies the biorthogonality conditions given as:

$$V_j \perp \tilde{W}_j, \quad \tilde{V}_j \perp W_j \quad \text{and} \quad W_j \perp \tilde{W}_{j'} \quad \text{for} \quad j \neq j' . \tag{2.30}$$

By definition, a scaling function and a mother wavelet satisfy the dilation relation and the wavelet relation, respectively. So we have:

$$\phi(t) = \sum_k h_k \phi(2t - k) \quad \text{and} \quad \psi(t) = \sum_k g_k \phi(2t - k) \tag{2.31}$$

and similarly

$$\tilde{\phi}(t) = \sum_k \tilde{h}_k \tilde{\phi}(2t - k) \quad \text{and} \quad \tilde{\psi}(t) = \sum_k \tilde{g}_k \tilde{\phi}(2t - k) . \tag{2.32}$$

The coefficients in the above equations can be obtained by taking the inner product with the appropriate dual function. For example,

$$h_k = \left\langle \tilde{\phi}_{1k}, \phi \right\rangle \quad \text{and} \quad g_k = \left\langle \tilde{\phi}_{1k}, \psi \right\rangle . \tag{2.33}$$

Note that in addition to ϕ and $\tilde{\phi}$, the roles of ψ and $\tilde{\psi}$ can also be interchanged.

CHAPTER 3

Formulation of the Multiresolution Frequency Domain Schemes

This chapter presents the derivation of the biorthogonal wavelet based Multiresolution Frequency Domain scheme for homogeneous problems. Also included are the derivation of the Finite Difference Frequency Domain method from MoM formulation and the selection criteria of the appropriate wavelet bases.

3.1 DERIVATION OF THE FINITE DIFFERENCE FREQUENCY DOMAIN SCHEME BY THE METHOD OF MOMENTS

It has been observed in [25] that the finite difference time domain scheme can be derived by applying method of moments [10] to Maxwell's curl equations while using pulse functions as a basis for the expansion of unknown fields. This observation and the fact that MoM can use any orthonormal set of functions as basis functions lead to the development of numerous multiresolution based time domain schemes [6, 7, 8, 9]. It will be shown here that the finite difference frequency domain technique can also be derived by the aid of the method of moments.

Similar to the procedure in [26], a one-dimensional, one-way wave equation is considered as a simple example:

$$\frac{\partial E(z)}{\partial z} = \frac{j\omega}{c} E(z). \tag{3.1}$$

This wave equation can be modeled by the FDFD method using the central difference approximation:

$$\frac{E_{k+1} - E_{k-1}}{2\Delta z} = \frac{j\omega}{c} E_k \tag{3.2}$$

where E_k corresponds to the electric field located at the grid node $z = k\Delta z$ and Δz is the distance between two neighboring grid nodes.

However, there is an alternative approach which will enable us to acquire the same result. This approach, which is based on the method of moments, provides the basis for the development of multiresolution frequency domain formulations.

Let us apply MoM to the wave equation (3.1). For the discretization of (3.1) by the method of moments, the electric field is expanded in terms of pulse functions:

$$E(z) = \sum_{k'} E_{k'} \phi_{k'}(z) .$$
(3.3)

$\phi_{k'}(z)$ is the pulse function defined as:

$$\phi_{k'}(z) = \begin{cases} 1, & k'\Delta z \leq z < (k'+1)\Delta z \\ 0, & \text{otherwise} \end{cases}$$
(3.4)

where k' is an integer.

Substituting (3.3) into (3.1) yields:

$$\sum_{k'} E_{k'} \frac{d\phi_{k'}(z)}{dz} = \frac{j\omega}{c} \sum_{k'} E_{k'} \phi_{k'}(z) .$$
(3.5)

Following the method of moments procedure, (3.5) is sampled by the complex conjugate of the testing functions. During the sampling procedure, the following integrals are utilized [26]:

$$\int_{-\infty}^{\infty} \frac{d\phi_{k'}(z)}{dz} \phi_k(z) dz = \frac{1}{2}(\delta_{k',k+1} - \delta_{k',k-1})$$
(3.6)

$$\int_{-\infty}^{\infty} \phi_{k'}(z) \phi_k(z) dz = \delta_{k',k} \Delta z$$
(3.7)

where $\delta_{k',k}$ is the Kronecker's delta described by:

$$\delta_{k',k} = \begin{cases} 1, & \text{if } k' = k \\ 0, & \text{otherwise} . \end{cases}$$
(3.8)

Sampling the left-hand side of (3.5) yields:

$$\int_{-\infty}^{\infty} \sum_k E_{k'} \frac{d\phi_{k'}(z)}{dz} \phi_k(z) dz = \sum_{k'} E_{k'} \int_{-\infty}^{\infty} \frac{d\phi_{k'}(z)}{dz} \phi_k(z) dz$$

$$= \frac{1}{2} \sum_{k'} E_{k'}(\delta_{k',k+1} - \delta_{k',k-1})$$
(3.9)

$$= \frac{E_{k+1} - E_{k-1}}{2}.$$

Similarly, sampling the right-hand side of (3.5) yields:

$$\int_{-\infty}^{\infty} \frac{j\omega}{c} \sum_{k'} E_{k'} \phi_{k'}(z)\phi_k(z)dz = \frac{j\omega}{c} \sum_{k'} E_{k'} \int_{-\infty}^{\infty} \phi_{k'}(z)\phi_k(z)dz$$

$$= \frac{j\omega}{c} \sum_{k'} E_{k'} \delta_{k',k} \Delta z \qquad (3.10)$$

$$= \frac{j\omega}{c} E_k \Delta z.$$

Finally, in view of (3.5), the right-hand sides of (3.9) and (3.10) are equal to each other so that

$$\frac{E_{k+1} - E_{k-1}}{2\Delta z} = \frac{j\omega}{c} E_k \qquad (3.11)$$

which reveals that the FDFD formulation given by (3.2) is obtained again. Therefore, it is acknowledged that discretizing the wave equation by FDFD and MoM is identical.

3.2   SELECTION OF THE APPROPRIATE WAVELET FAMILY

In the previous section, it has been proved that the FDFD formulation is a special case of the method of moments where the unknown field functions are expanded in terms of pulse functions. The pulse function is also known as the scaling function of the Haar wavelet base. Thus, the FDFD technique can also be considered a multiresolution analysis based scheme.

The Haar wavelet is not a smooth function; hence approximating smooth functions like electromagnetic fields with this wavelet family is inefficient. This is the fundamental reason for the main limitation of FDFD scheme: numerical dispersion. It is a well-known fact that the method of moments permits the use of any set of orthonormal functions as a basis of expansion; hence it is possible to replace the Haar scaling function with the scaling function of another wavelet base that can approximate electromagnetic fields more efficiently. The performance characteristics of the resulting multiresolution scheme will be determined by the new wavelet base, so the choice of wavelet family is critical.

In order to develop an efficient finite difference formulation based on multiresolution analysis, one should carefully choose the appropriate wavelet family from an ever-increasing number of wavelets available. The appropriate wavelet base should have certain properties, such as compact support, symmetry, interpolation property, regularity (smoothness), and maximum number of vanishing moments.

Probably the most useful class of scaling functions is those that have compact support or finite support. A function has finite support if it is identically zero outside of a finite interval and is said to have compact support if this interval is narrow. The Haar scaling function is a good example of a compactly supported function. The support of the wavelet basis is directly related to the number of terms in the update equations. Since a great number of terms in each update equation will increase the computational burden and complicate the coding process, it is preferable to have a compactly supported wavelet base.

Symmetric wavelet functions will ensure the symmetry of the formulation and in return will simplify the modeling of symmetry and boundary planes [27].

Compact support and symmetry are incompatible aspects of orthogonal wavelet systems, with the Haar wavelets being the exception. Unfortunately, Haar wavelets are not smooth, so they cannot approximate a smooth function effectively. However, a biorthogonal wavelet base can sustain smooth, compact, and symmetric wavelets, which is the main reason for the choice of biorthogonal wavelets over orthogonal ones.

In a wavelet expansion, the field value at a certain point is generally reconstructed by a weighted sum of related neighboring basis function coefficients, which requires a complex reconstruction algorithm. However, the reconstruction algorithm is not essential if the basis function satisfies the interpolation property, as in such cases the basis function coefficients represent the field components at the corresponding position of the grid [28]. Thus, a basis function equipped with the interpolation property eliminates the use of the weighted sum and thus yields a more efficient MRFD algorithm.

Two of the most desired properties of a wavelet family are high regularity and maximum number of vanishing moments, owing to their great effect on how well the wavelet expansion approximates a smooth function. Unfortunately, wavelets do not accommodate both high regularity and high number of vanishing moments concurrently and it is not always clear which property is more important [24]. Biorthogonal wavelets accommodate two scaling functions, which may have different regularity properties. It is mentioned in [24] that it is beneficial to reserve the high regularity to the synthesis scaling function and maximum vanishing moments to the analysis scaling function.

The Battle-Lemarie wavelet functions [29, 30] are symmetric and smooth. However, these wavelet functions do not have finite support. This results in an infinite number of terms in MRFD update equations. Fortunately, these wavelets decay exponentially, so truncating the MRFD scheme with respect to space is a possible solution with negligible error.

The Cohen-Daubechies-Feauveau (CDF) family of wavelets [31] are symmetric biorthogonal wavelets that accommodate the interpolation property and compact support. The dual scaling function of this family has high regularity whereas the scaling function has a large number of vanishing moments. Considering the desired attributes mentioned above, the CDF family of wavelets, in particular the CDF(2,2) wavelet, is mainly adopted for this work due to its minimal support. Another argument worth mentioning is the successful application of these wavelets in time-domain multiresolution analysis schemes [6, 28, 32, 33, 34]. The wavelet and scaling functions of the considered wavelet families are sketched in Figures 3.1–3.5 by the aid of the Matlab Wavelet Toolbox.

3.3   DERIVATION OF THE MULTIRESOLUTION FREQUENCY DOMAIN SCHEME

In this section, derivation of the biorthogonal wavelet based multiresolution frequency domain method for lossless homogeneous media is presented. The procedure followed is similar to the one

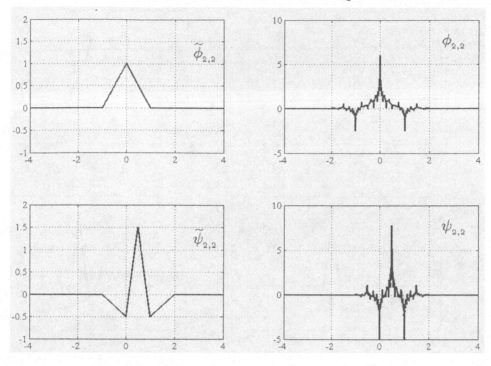

Figure 3.1: Scaling and wavelet functions of the CDF(2,2) wavelet base.

presented in Section 3.1, however this time the fields are expanded in terms of biorthogonal wavelets instead of the Haar wavelets, and Maxwell's time-harmonic curl equations are discretized instead of the simple 1D wave equation. Thus, a more general formulation that can model arbitrary 3D structures is developed. The derivation of orthogonal wavelet-based MRFD schemes is similar to the derivation presented below.

We consider a lossless homogeneous medium. In this medium, electric and magnetic fields are governed by the following time-harmonic Maxwell's equations:

$$\begin{bmatrix} 0 & -\partial/\partial z & \partial/\partial y \\ \partial/\partial z & 0 & -\partial/\partial x \\ -\partial/\partial y & \partial/\partial x & 0 \end{bmatrix} \begin{bmatrix} H_x \\ H_y \\ H_z \end{bmatrix} = j\omega \begin{bmatrix} \varepsilon_x & 0 & 0 \\ 0 & \varepsilon_y & 0 \\ 0 & 0 & \varepsilon_z \end{bmatrix} \begin{bmatrix} E_x \\ E_y \\ E_z \end{bmatrix} \qquad (3.12a)$$

$$\begin{bmatrix} 0 & -\partial/\partial z & \partial/\partial y \\ \partial/\partial z & 0 & -\partial/\partial x \\ -\partial/\partial y & \partial/\partial x & 0 \end{bmatrix} \begin{bmatrix} E_x \\ E_y \\ E_z \end{bmatrix} = -j\omega \begin{bmatrix} \mu_x & 0 & 0 \\ 0 & \mu_y & 0 \\ 0 & 0 & \mu_z \end{bmatrix} \begin{bmatrix} H_x \\ H_y \\ H_z \end{bmatrix}. \qquad (3.12b)$$

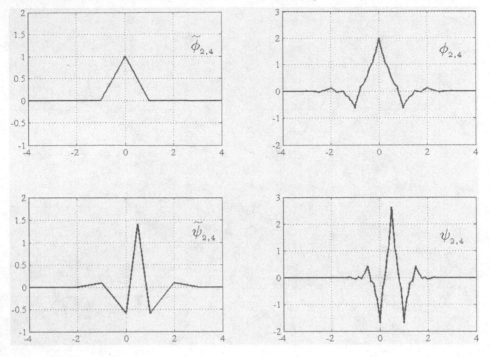

Figure 3.2: Scaling and wavelet functions of the CDF(2,4) wavelet base.

In the Cartesian coordinate system, these curl equations lead to six scalar equations:

$$j\omega\varepsilon_x E_x(x, y, z) = \frac{\partial H_z(x, y, z)}{\partial y} - \frac{\partial H_y(x, y, z)}{\partial z} \tag{3.13a}$$

$$j\omega\varepsilon_y E_y(x, y, z) = \frac{\partial H_x(x, y, z)}{\partial z} - \frac{\partial H_z(x, y, z)}{\partial x} \tag{3.13b}$$

$$j\omega\varepsilon_z E_z(x, y, z) = \frac{\partial H_y(x, y, z)}{\partial x} - \frac{\partial H_x(x, y, z)}{\partial y} \tag{3.13c}$$

$$j\omega\mu_x H_x(x, y, z) = -\frac{\partial E_z(x, y, z)}{\partial y} + \frac{\partial E_y(x, y, z)}{\partial z} \tag{3.13d}$$

$$j\omega\mu_y H_y(x, y, z) = -\frac{\partial E_x(x, y, z)}{\partial z} + \frac{\partial E_z(x, y, z)}{\partial x} \tag{3.13e}$$

$$j\omega\mu_z H_z(x, y, z) = -\frac{\partial E_y(x, y, z)}{\partial x} + \frac{\partial E_x(x, y, z)}{\partial y}. \tag{3.13f}$$

For the discretization of these equations by the method of moments, the field components should be first expanded in terms of the basis functions. Biorthogonal wavelets accommodate two

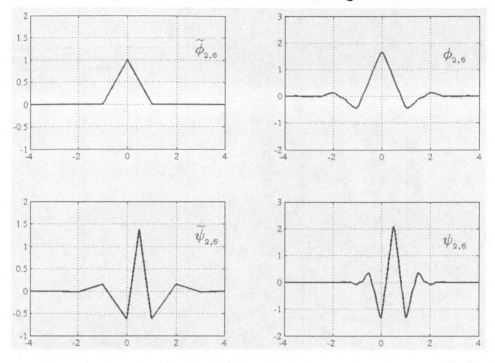

Figure 3.3: Scaling and wavelet functions of the CDF(2,6) wavelet base.

scaling functions, either of which can be picked as a basis of expansion. However, as mentioned previously, it is beneficial to reserve the high regularity to the synthesis scaling function and maximum vanishing moments to the analysis scaling function. The dual scaling functions of CDF wavelet bases ($\tilde{\phi}(x)$) employ higher regularity, so they will be used for the expansion of the unknown fields.

$$E_x(x, y, z) = \sum_{i', j', k'} E_x(i', j', k')\tilde{\phi}_{i'+1/2}(x)\tilde{\phi}_{j'}(y)\tilde{\phi}_{k'}(z) \tag{3.14a}$$

$$E_y(x, y, z) = \sum_{i', j', k'} E_y(i', j', k')\tilde{\phi}_{i'}(x)\tilde{\phi}_{j'+1/2}(y)\tilde{\phi}_{k'}(z) \tag{3.14b}$$

$$E_z(x, y, z) = \sum_{i', j', k'} E_z(i', j', k')\tilde{\phi}_{i'}(x)\tilde{\phi}_{j'}(y)\tilde{\phi}_{k'+1/2}(z) \tag{3.14c}$$

$$H_x(x, y, z) = \sum_{i', j', k'} H_x(i', j', k')\tilde{\phi}_{i'}(x)\tilde{\phi}_{j'+1/2}(y)\tilde{\phi}_{k'+1/2}(z) \tag{3.14d}$$

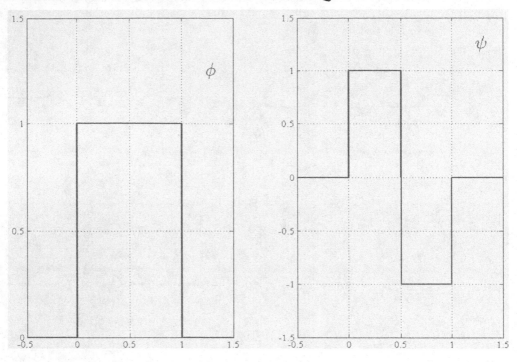

Figure 3.4: Scaling and wavelet functions of the Haar wavelet base.

$$H_y(x, y, z) = \sum_{i',j',k'} H_y(i', j', k')\tilde{\phi}_{i'+1/2}(x)\tilde{\phi}_{j'}(y)\tilde{\phi}_{k'+1/2}(z) \tag{3.14e}$$

$$H_z(x, y, z) = \sum_{i',j',k'} H_z(i', j', k')\tilde{\phi}_{i'+1/2}(x)\tilde{\phi}_{j'+1/2}(y)\tilde{\phi}_{k'}(z) . \tag{3.14f}$$

Here, the indexes i', j', k' indicate the discrete space lattice related to the space grid through $x = i'\Delta x$, $y = j'\Delta y$ and $z = k'\Delta z$. This notation is basically the same as Yee's notation. There is a strong relation between these field expansions and the Yee cell. The scaling functions are placed such that they are centered at the location of the corresponding field component on the Yee cell. Similar to the FDFD notation, half integral values of i, j, and k in the expansion coefficients (such as $E_x(i, j, k)$, $H_x(i, j, k)$) are represented by integer indices, as explained in Table 2.2.

The function $\tilde{\phi}_n(x)$ is the scaled and dilated dual scaling function ($\tilde{\phi}(x)$) defined as:

$$\tilde{\phi}_n(x) = \tilde{\phi}\left(\frac{x - n\Delta x}{\Delta x}\right) . \tag{3.15}$$

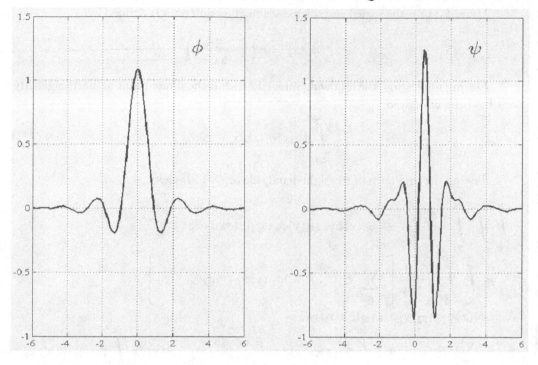

Figure 3.5: Scaling and wavelet functions of the Battle-Lemarie wavelet base.

For the discretization of Eq. (3.12b), one of the corresponding scalar equations, (3.13d), is considered as an example. First, the field expansions are inserted into (3.13d) and both sides are tested with the scaling function according to MoM.

Testing the left-hand side of (3.13d) yields:

$$j\omega\mu_x \int\limits_{-\infty}^{\infty} \int\limits_{-\infty}^{\infty} \int\limits_{-\infty}^{\infty} H_x(x, y, z)\phi_i(x)\phi_{j+1/2}(y)\phi_{k+1/2}(z)dxdydz$$

$$= j\omega\mu_x \int\limits_{-\infty}^{\infty} \int\limits_{-\infty}^{\infty} \int\limits_{-\infty}^{\infty} \left[\sum_{i',j',k'} H_x(i', j', k')\tilde{\phi}_{i'}(x)\tilde{\phi}_{j'+1/2}(y)\tilde{\phi}_{k'+1/2}(z) \right]$$

$$\phi_i(x)\phi_{j+1/2}(y)\phi_{k+1/2}(z)dxdydz$$

$$= j\omega\mu_x \sum_{i',j',k'} H_x(i', j', k')\Delta x\delta_{i',i}\Delta y\delta_{j'+1/2,j+1/2}\Delta z\delta_{k'+1/2,k+1/2}$$

$$= j\omega\mu_x H_x(i, j, k)\Delta x\Delta y\Delta z.$$

(3.16)

Here, $\phi_n(x)$ is the scaled and dilated scaling function ($\phi(x)$) defined by:

$$\phi_n(x) = \phi\left(\frac{x - n\Delta x}{\Delta x}\right). \tag{3.17}$$

During the testing, the following integral which is the direct result of biorthogonality conditions has been employed:

$$\int_{-\infty}^{\infty} \tilde{\phi}_{i'}(x)\phi_i(x)dx = \Delta x \delta_{i',i}. \tag{3.18}$$

Testing the first term of the right-hand side of (3.13d) yields:

$$\int_{-\infty}^{\infty}\int_{-\infty}^{\infty}\int_{-\infty}^{\infty} \frac{\partial E_z(x, y, z)}{\partial y}\phi_i(x)\phi_{j+1/2}(y)\phi_{k+1/2}(z)dxdydz$$

$$= \int_{-\infty}^{\infty}\int_{-\infty}^{\infty}\int_{-\infty}^{\infty} \frac{\partial}{\partial y}\left[\sum_{i',j',k'} E_z(i', j', k')\tilde{\phi}_{i'}(x)\tilde{\phi}_{j'}(y)\tilde{\phi}_{k'+1/2}(z)\right]$$

$$\phi_i(x)\phi_{j+1/2}(y)\phi_{k+1/2}(z)dxdydz \tag{3.19}$$

$$= \sum_{i',j',k'} E_z(i', j', k') \int_{-\infty}^{\infty} \tilde{\phi}_{i'}(x)\phi_i(x)dx \int_{-\infty}^{\infty} \frac{\partial\tilde{\phi}_{j'}(y)}{\partial y}\phi_{j+1/2}(y)dy \int_{-\infty}^{\infty} \tilde{\phi}_{k'+1/2}(z)\phi_{k+1/2}(z)dz$$

$$= \sum_{i',j',k'} E_z(i', j', k')\Delta x \delta_{i',i} \Delta z \delta_{k'+1/2,k+1/2}$$

$$\left[\cdots + a(-2)\delta_{j',j-2} + a(-1)\delta_{j',j-1} + a(0)\delta_{j',j}\right.$$
$$\left. + a(1)\delta_{j',j+1} + a(2)\delta_{j',j+2} + a(3)\delta_{j',j+3} + \cdots\right]$$

$$= \Delta x \Delta z \sum_{l=-n_a+1}^{n_a} a(l)E_z(i, j+l, k)$$

In order to establish the above equations, biorthogonal relation (3.18) and the following integral were employed [35]:

$$a(l) = \int_{-\infty}^{\infty} \frac{\partial\tilde{\phi}_{i+l}(x)}{\partial x}\phi_{i+1/2}(x)dx. \tag{3.20}$$

Similarly, testing the second term of the right-hand side of (3.13d) yields:

$$\int_{-\infty}^{\infty}\int_{-\infty}^{\infty}\int_{-\infty}^{\infty} \frac{\partial E_y(x,y,z)}{\partial z}\phi_i(x)\phi_{j+1/2}(y)\phi_{k+1/2}(z)dxdydz$$

$$= \Delta x \Delta y \sum_{l=-n_a+1}^{n_a} a(l)E_y(i, j, k+l). \tag{3.21}$$

Here, n_a is the number of non-zero $a(l)$ coefficients and is called the stencil size of the wavelet base. The value of the stencil size depends solely on the support of the wavelet base. A wide support means a big stencil size and many elements in the update equations, whereas a narrow support means few elements.

The $a(l)$ coefficients of different wavelet families are presented in Table 3.1 [6]. Note that the Battle-Lemarie wavelet family, which occupies the last column of the table, has infinite support, and thus the number of coefficients is also infinite. However, the values of the coefficients decay exponentially and get negligible beyond the 9th coefficient.

If the wavelet base functions are symmetric, then the symmetry relation given by $a(-l) = -a(l+1)$ [36] holds. Using the symmetry relation, and equating (3.16), (3.19) and (3.21), (3.13d) leads to the MRFD update equation:

$$j\omega\mu_x H_x(i,j,k) = \left(\frac{\sum\limits_{l=1}^{n_a} a(l)[E_y(i,j,k+l)-E_y(i,j,k-l+1)]}{\Delta z} - \frac{\sum\limits_{l=1}^{n_a} a(l)[E_z(i,j+l,k)-E_z(i,j-l+1,k)]}{\Delta y} \right). \tag{3.22}$$

Table 3.1: $a(l)$ coefficients of various wavelets					
l	Haar	CDF(2,2)	CDF(2,4)	CDF(2,6)	Battle-Lemarie
1	1	1.2291667	1.2918134	1.3110337	1.2918462
2	0	-0.0937500	-0.1371348	-0.1560124	-0.1560761
3	0	0.0104167	0.0287617	0.0419962	0.0596391
4	0	0	-0.0034701	-0.0086543	-0.0293099
5	0	0	0.0000080	0.0008308	0.0153716
6	0	0	0	0.0000109	-0.0081892
7	0	0	0	-0.0000000	0.0043788
8	0	0	0	0	-0.0023433
9	0	0	0	0	0.0012542

Maxwell's curl equation (3.12a) is discretized in a similar fashion. The scalar equation (3.13a) is considered as an example. Again, the field expansions are inserted into the scalar equation and both sides are tested with the scaling function according to MoM. First, the left-hand side of (3.13a) is tested:

$$
j\omega\varepsilon_x \int_{-\infty}^{\infty}\int_{-\infty}^{\infty}\int_{-\infty}^{\infty} E_x(x,y,z)\phi_{i+1/2}(x)\phi_j(y)\phi_k(z)dxdydz
$$

$$
= j\omega\varepsilon_x \int_{-\infty}^{\infty}\int_{-\infty}^{\infty}\int_{-\infty}^{\infty} \left[\sum_{i',j',k'} E_x(i',j',k')\tilde{\phi}_{i'+1/2}(x)\tilde{\phi}_{j'}(y)\tilde{\phi}_{k'}(z) \right] \phi_{i+1/2}(x)\phi_j(y)\phi_k(z)dxdydz
$$

$$
= j\omega\varepsilon_x \sum_{i',j',k'} E_x(i',j',k').\Delta x\delta_{i'+1/2,i+1/2}.\Delta y\delta_{j',j}.\Delta z\delta_{k',k}
$$

$$
= j\omega\varepsilon_x E_x(i,j,k)\Delta x\Delta y\Delta z. \tag{3.23}
$$

Then the first term of the right-hand side of (3.13a) is tested:

$$
\int_{-\infty}^{\infty}\int_{-\infty}^{\infty}\int_{-\infty}^{\infty} \frac{\partial H_z(x,y,z)}{\partial y}\phi_{i+1/2}(x)\phi_j(y)\phi_k(z)dxdydz
$$

$$
= \int_{-\infty}^{\infty}\int_{-\infty}^{\infty}\int_{-\infty}^{\infty} \frac{\partial}{\partial y}\left[\sum_{i,j,k} H_z(i',j',k')\tilde{\phi}_{i'+1/2}(x)\tilde{\phi}_{j'+1/2}(y)\tilde{\phi}_{k'}(z) \right]
$$

$$
\phi_{i+1/2}(x)\phi_j(y)\phi_k(z)dxdydz
$$

$$
= \sum_{i',j',k'} H_z(i',j',k') \int_{-\infty}^{\infty} \tilde{\phi}_{i'+1/2}(x)\phi_{i+1/2}(x)dx \tag{3.24}
$$

$$
\int_{-\infty}^{\infty} \frac{\partial\tilde{\phi}_{j'+1/2}(y)}{\partial y}\phi_j(y)dy \int_{-\infty}^{\infty} \tilde{\phi}_{k'}(z)\phi_k(z)dz
$$

$$
= \sum_{i',j',k'} H_z(i',j',k').\Delta x\delta_{i'+1/2,i+1/2}.\Delta z\delta_{k',k} .
$$

$$
[\ldots + a(-2)\delta_{j',j-3} + a(-1)\delta_{j',j-2} + a(0)\delta_{j',j-1}
$$
$$
+ a(1)\delta_{j',j} + a(2)\delta_{j',j+1} + a(3)\delta_{j',j+2} + \ldots]
$$

$$
= \Delta x\Delta z \sum_{l=-n_a+1}^{n_a} a(l)H_z(i,j+l-1,k).
$$

Finally, the second term of the right-hand side of (3.13a) can similarly be sampled to yield:

$$
\int_{-\infty}^{\infty} \int_{-\infty}^{\infty} \int_{-\infty}^{\infty} \frac{\partial H_y(x, y, z)}{\partial z} \phi_{i+1/2}(x)\phi_j(y)\phi_k(z)\,dx\,dy\,dz
$$

$$
= \Delta x \Delta y \sum_{l=-n_a+1}^{n_a} a(l) H_y(i, j, k+l-1). \tag{3.25}
$$

Using the symmetry relation and use of (3.23)–(3.25) in the tested (3.13a) results in the MRFD update equation:

$$
j\omega\varepsilon_x E_x(i, j, k) = \left(\begin{array}{c} \dfrac{\sum_{l=1}^{n_a} a(l)\left[H_z(i, j+l-1, k) - H_z(i, j-l, k) \right]}{\Delta y} \\[1em] -\dfrac{\sum_{l=1}^{n_a} a(l)\left[H_y(i, j, k+l-1) - H_y(i, j, k-l) \right]}{\Delta z} \end{array} \right). \tag{3.26}
$$

During the sampling process, the following integral is employed:

$$
a(l) = \int_{-\infty}^{\infty} \frac{\partial \tilde{\phi}_{i+l-1/2}(x)}{\partial x} \phi_i(x)\,dx. \tag{3.27}
$$

Two scalar equations, each corresponding to one of Maxwell's time-harmonic curl equations, were discretized by using a MRFD scheme. The derivation of the remaining four update equations can be executed similarly.

Since the MRFD formulation is derived for homogeneous mediums, no relationship was established between the material properties (ε, μ) and the discretization lattice. We would like to model inhomogeneous mediums with the MRFD formulation. For the sampling of material parameters, the point-wise sampling technique of FDFD is adopted. The positions of the material parameters are associated with the field components: values of ε are associated with $\vec{E}$ field components whereas values of μ are associated with $\vec{H}$ field components. The spatial location of the material parameters on the unit cell and the corresponding numbering scheme are listed in Table 2.2.

The pointwise sampling is mathematically valid only for the homogeneous materials but the error introduced by modeling inhomogeneous mediums by pointwise sampling can be neglected [28], especially if the support of the wavelet base is minimal. With the point-wise sampling applied, the complete set of update equations can be listed as:

$$j\omega\varepsilon_x(i,j,k)E_x(i,j,k) = \left(\frac{\sum\limits_{l=1}^{n_a} a(l)\left[H_z(i,j+l-1,k)-H_z(i,j-l,k)\right]}{\Delta y} - \frac{\sum\limits_{l=1}^{n_a} a(l)\left[H_y(i,j,k+l-1)-H_y(i,j,k-l)\right]}{\Delta z} \right) \tag{3.28a}$$

$$j\omega\varepsilon_y(i,j,k)E_y(i,j,k) = \left(\frac{\sum\limits_{l=1}^{n_a} a(l)\left[H_x(i,j,k+l-1)-H_x(i,j,k-l)\right]}{\Delta z} - \frac{\sum\limits_{l=1}^{n_a} a(l)\left[H_z(i+l-1,j,k)-H_z(i-l,j,k)\right]}{\Delta x} \right) \tag{3.28b}$$

$$j\omega\varepsilon_z(i,j,k)E_z(i,j,k) = \left(\frac{\sum\limits_{l=1}^{n_a} a(l)\left[H_y(i+l-1,j,k)-H_y(i-l,j,k)\right]}{\Delta x} - \frac{\sum\limits_{l=1}^{n_a} a(l)\left[H_x(i,j+l-1,k)-H_x(i,j-l,k)\right]}{\Delta y} \right) \tag{3.28c}$$

$$j\omega\mu_x(i,j,k)H_x(i,j,k) = \left(\frac{\sum\limits_{l=1}^{n_a} a(l)\left[E_y(i,j,k+l)-E_y(i,j,k-l+1)\right]}{\Delta z} - \frac{\sum\limits_{l=1}^{n_a} a(l)\left[E_z(i,j+l,k)-E_z(i,j-l+1,k)\right]}{\Delta y} \right) \tag{3.28d}$$

$$j\omega\mu_y(i,j,k)H_y(i,j,k) = \left(\frac{\sum\limits_{l=1}^{n_a} a(l)\left[E_z(i+l,j,k)-E_z(i-l+1,j,k)\right]}{\Delta x} - \frac{\sum\limits_{l=1}^{n_a} a(l)\left[E_x(i,j,k+l)-E_x(i,j,k-l+1)\right]}{\Delta z} \right) \tag{3.28e}$$

$$j\omega\mu_z(i,j,k)H_z(i,j,k) = \left(\frac{\sum\limits_{l=1}^{n_a} a(l)\left[E_x(i,j+l,k)-E_x(i,j-l+1,k)\right]}{\Delta y} - \frac{\sum\limits_{l=1}^{n_a} a(l)\left[E_y(i+l,j,k)-E_y(i-l+1,j,k)\right]}{\Delta x} \right). \tag{3.28f}$$

For the purpose of comparison of the FDFD and MRFD schemes, update equations for the E_x component using these two techniques are provided below, respectively. One can easily conclude that the FDFD formulation is the special case of the MRFD formulation with $a(1)=1$ and $n_a=1$.

$$j\omega\varepsilon_x(i,j,k)E_x(i,j,k) = \left(\frac{\left[H_z(i,j,k)-H_z(i,j-1,k)\right]}{\Delta y} - \frac{\left[H_y(i,j,k)-H_y(i,j,k-1)\right]}{\Delta z} \right) \tag{3.29}$$

$$j\omega\varepsilon_x(i,j,k)E_x(i,j,k) = \left(\frac{\sum\limits_{l=1}^{n_a} a(l)\left[H_z(i,j+l-1,k)-H_z(i,j-l,k)\right]}{\Delta y} - \frac{\sum\limits_{l=1}^{n_a} a(l)\left[H_y(i,j,k+l-1)-H_y(i,j,k-l)\right]}{\Delta z} \right). \tag{3.30}$$

CHAPTER 4

Application of MRFD Formulation to Closed Space Structures

In order to verify and demonstrate the efficiency of the new MRFD schemes, one, two, and three-dimensional closed space problems are considered in this chapter. The structures that are analyzed in this chapter are all assumed to be enclosed by a perfect electric conductor (PEC) boundary; hence the name "closed space structures." For the purpose of supporting the claim that MRFD formulation is inherently more efficient than the FDFD scheme, the closed space problems are solved with the aid of both methods and the numerical results are compared. The treatment of PEC boundary conditions in the context of the image principle and the multiple image technique is also addressed.

4.1   1D APPLICATION: THE FABRY-PEROT RESONATOR

The Fabry-Perot resonator structure is picked as an appropriate example of a one-dimensional closed space problem. A Fabry-Perot resonator is a resonant cavity formed by two parallel metal plates separated by a medium such as air or dielectric. A simple depiction of this structure is shown in Figure 4.1.

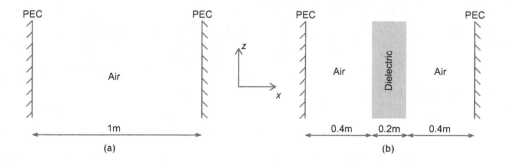

Figure 4.1: One-dimensional cavities.

When two air-separated metal plates are aligned perfectly parallel to each other, they can support a TEM standing wave described by:

$$E_z = E_0 \sin k_0 x \tag{4.1a}$$

$$H_y = -\frac{j E_0}{\eta_0} \cos k_0 x \tag{4.1b}$$

These standing waves resonate at frequencies:

$$f_n = \frac{n}{2d \sqrt{\mu_0 \varepsilon_0}} \quad for \ n = 1, 2, 3, \ldots. \tag{4.2}$$

where d is the distance between the two plates and n is the mode number.

FORMULATION

Since the standing wave between the plates is only a one-dimensional function, the structure can be reduced to a 1D problem by assuming $\partial / \partial y = 0$ and $\partial / \partial z = 0$. In this case, Maxwell's vector curl equations can be reduced to two scalar equations:

$$\frac{\partial E_z}{\partial x} - j\omega\mu_y H_y = 0 \tag{4.3a}$$

$$\frac{\partial H_y}{\partial x} - j\omega\varepsilon_z E_z = 0. \tag{4.3b}$$

Using the one-dimensional grid shown in Figure 4.2, the MRFD update equations for (4.3) can be simplified from (3.28c) and (3.28e) as:

$$\omega E_z(i) = \frac{\sum_{l=1}^{n_a} a(l) \left[H_y(i + l - 1) - H_y(i - l) \right]}{j\varepsilon(i)\Delta x} \tag{4.4a}$$

$$\omega H_y(i) = \frac{\sum_{l=1}^{n_a} a(l) \left[E_z(i + l) - E_z(i - l + 1) \right]}{j\mu(i)\Delta x}. \tag{4.4b}$$

Equation (4.4) can be used to produce an eigenvalue problem:

$$\omega . x = A . x. \tag{4.5}$$

Here, A is the sparse coefficient matrix and x is the unknown field vector. The eigenvalues of A are the resonant frequencies of the structure. The eigenvectors of A provide the corresponding electromagnetic fields.

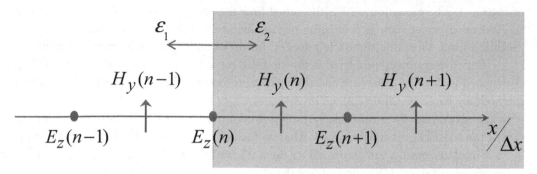

Figure 4.2: One-dimensional grid.

TREATMENT OF THE BOUNDARIES AND DIELECTRIC INTERFACES

At the PEC boundary, the tangential electric and normal magnetic fields are equal to zero. A conventional FDFD scheme can model the PEC boundary simply by setting the coefficient of tangential electric and normal magnetic fields to zero, owing to the localized scaling function. Multiresolution schemes, however, are non-localized in nature because of the wider support of scaling functions, so they cannot support localized boundary conditions. This aspect of multiresolution schemes is caused by the fact that the update equations of the grid nodes in the vicinity of the boundary include the field components outside of the computational domain.

For the truncation of the PEC boundary, the image principle [9] is adopted. In this method, the scaling coefficients of the fields on the boundary are set to zero and even or odd symmetry is used to generate the fields outside the computational domain. For the 1D case study at hand, the electric and magnetic fields are both tangential to the boundary, so the electric field and magnetic field components are forced to have odd and even symmetry across the PEC boundary, respectively. For example, if the PEC boundary coincides with MRFD lattice at the grid point $x = 0$, than $E(-1) = -E(1)$ and $H(-1) = H(0)$.

At the interface between two materials, averaging of material parameters is used. An interface between two mediums is depicted in Figure 4.2. For this interface, averaging of dielectric constant requires $\varepsilon(n-1) = \varepsilon_1$, $\varepsilon(n) = \frac{\varepsilon_1 + \varepsilon_2}{2}$, and $\varepsilon(n+1) = \varepsilon_2$.

NUMERICAL RESULTS

In order to compare the efficiency and accuracy of FDFD and MRFD schemes, the resonant frequencies of the one-dimensional resonator structures shown in Figure 4.1 are calculated with FDFD and various MRFD methods.

The source codes are written in Matlab and executed using a laptop PC equipped with a Pentium M processor at 1.6 GHz and 1 GB of memory. The coefficient matrices are stored in sparse matrices and Matlab function *eigs* is employed during the calculation of eigenvalues and eigenvectors of matrix A. The execution time is the actual time required to fill and solve the coefficient matrix A. Additional steps, such as processing or visualizing the calculated data, are not included in the execution time. The same approach is applied for the examples in the following sections.

The first example considered is the air-filled resonator shown in Figure 4.1a. This structure is analyzed with finite difference and three different multiresolution methods. The multiresolution methods are based on CDF(2,2), CDF(2,4), and CDF(2,6) wavelets. MRFD analyses are executed with a mesh that has three times lower resolution compared to the FDFD lattice. In this case, the cell size of the MRFD grid equals $\lambda_{min}/4$, whereas the cell size of the FDFD grid is equal to $\lambda_{min}/12$.

Simulated results are compared to the analytical values based on (4.2), and the results are summarized in Table 4.1. Results indicate that, for a one-dimensional homogeneous case, the MRFD methods manage to perform like the of FDFD method in terms of accuracy while reducing the grid resolution by a factor of 3:1. Memory requirements, simulation time, and accuracy for different MRFD techniques vary based on the stencil size of the wavelets that are used. MRFD methods with higher stencil sizes achieve higher accuracy at the expense of higher memory requirements and simulation times.

Table 4.1: Calculated resonance frequencies, calculation error, and simulation time of the air-filled resonator

Wavelet Function	Analytic Resonance Frequency (MHz)	Calculated Resonance Frequency (MHz)	Error (%)	Number of Cells	Matrix Size (byte)	Simulation Time (msec)
Haar (FDFD)	149.896	149.788	0.07	24	2112	343
	299.792	298.930	0.29			
	449.689	446.791	0.64			
	599.585	592.732	1.14			
CDF (2,2)	149.896	149.916	0.01	8	1684	250
	299.792	300.319	0.18			
	449.689	452.673	0.66			
	599.585	607.295	1.29			
CDF (2,4)	149.896	149.896	0.00	8	2084	271
	299.792	299.951	0.05			
	449.689	451.708	0.45			
	599.585	609.684	1.68			
CDF (2,6)	149.896	149.907	0.01	8	2284	297
	299.792	299.815	0.01			
	449.689	450.555	0.19			
	599.585	605.953	1.06			

More comprehensive conclusions can be attained by analyzing the calculation error with respect to discretization points per wavelength. For this purpose, the higher order modes of the structure are simulated using a much higher grid density with a cell size of 1 cm. The errors introduced by MRFD methods as a function of number of cells per wavelength are presented in Figure 4.3. For comparison, FDFD results are also provided.

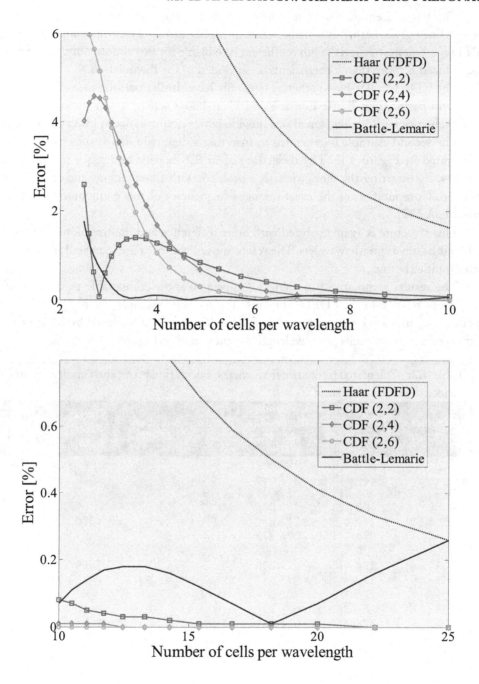

Figure 4.3: Error comparison of MRFD and FDFD methods, 1D uniform cavity.

The Battle-Lemarie-based scheme generally has the best accuracy for medium sized grids, however finer grids with this scheme result in higher errors. The truncation of the higher order MRFD coefficients beyond the 9th coefficient is to blame for this abnormality. This error is expected to be reduced if more than 9 coefficients are included in the formulation.

The CDF wavelet-based schemes generally have similar performance at high resolution, and all perform better than Battle-Lemarie and Haar-based schemes. On the other hand, the CDF wavelet schemes with higher stencil size provide better performance at lower resolutions.

The second example is generated by inserting a dielectric slab at the center of the resonator, as illustrated in Figure 4.1b. The dielectric slab is 0.2 m wide and has a relative dielectric constant of 4. By inserting the dielectric slab, a problem with three regions and two discontinuities is generated. Computation of the exact resonant frequencies of this configuration is summarized in Appendix A [37].

This structure is again analyzed with finite difference and multiresolution methods based on CDF and Battle-Lemarie wavelets. The resolutions of the grids are increased in order to model the discontinuities better.

The results summarized in Table 4.2 appear to show comparable percentage of error with time savings relative to the FDFD. Similar to the previous example, the higher order modes of the structure are simulated using a cell size of 1 cm. The errors introduced by MRFD methods as a function of number of cells per wavelength are presented in Figure 4.4.

Table 4.2: Calculated resonance frequencies, calculation error, and simulation time of the dielectric loaded resonator

Wavelet Function	Analytic Resonance Frequency (MHz)	Calculated Resonance Frequency (MHz)	Error (%)	Number of Cells	Matrix Size (byte)	Simulation Time (msec)
Haar (FDFD)	100.326	100.349	0.02	30	2660	625
	274.415	272.996	0.52			
	374.741	373.009	0.46			
CDF (2,2)	100.326	100.504	0.18	20	4580	406
	274.415	272.877	0.56			
	374.741	375.273	0.14			
CDF (2,4)	100.326	100.563	0.24	20	6980	436
	274.415	272.808	0.59			
	374.741	375.081	0.09			
CDF (2,6)	100.326	100.514	0.19	20	8180	472
	274.415	272.796	0.59			
	374.741	374.762	0.01			
Battle-Lemarie	100.326	101.045	0.72	20	10980	516
	274.415	272.969	0.53			
	374.741	374.206	0.14			

Results are similar to the homogeneous case except that the CDF(2,2) wavelet based formulation provides the best cost/performance ratio. This is mainly attributed to fact that this wavelet

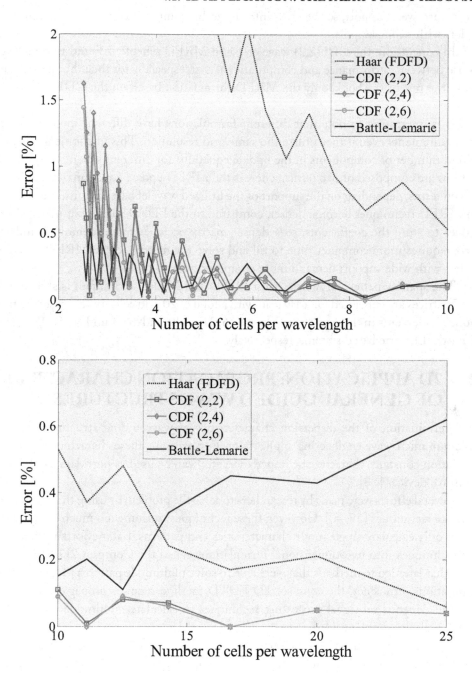

Figure 4.4: Error comparison of MRFD and FDFD methods, 1D non-uniform cavity.

has the narrowest support, so the error introduced by pointwise sampling is minimal compared to wavelets with wider support.

In conclusion, the CDF(2,2) wavelet-based MRFD scheme is found to offer the best compromise between performance and computational cost, especially for the inhomogeneous problems. During the rest of the book, only the MRFD formulation based on the CDF(2,2) wavelet is considered.

It is interesting to note that different formulations have different memory and simulation time requirements even if they utilize the same grid resolution. This phenomenon is caused by the different number of components in the update equations for different schemes. The FDFD update equations are composed of two terms, whereas the MRFD update equations are composed of six to eighteen terms, depending on the support of the utilized wavelet base. In turn, the coefficient matrix A in MRFD techniques become denser, compared to the FDFD coefficient matrix. We use sparse matrices to store the coefficients, so a denser matrix occupies more memory. Similarly, a denser matrix requires more computer time to fill and solve the matrix, so the MRFD schemes based on wavelets with wide support demand more computer time.

The sparsity pattern of matrix A for FDFD and MRFD techniques is shown in Figure 4.5. These matrices are generated for a 1D grid with a total of 20 grid cells. In this case, the total numbers of nonzero elements in the sparse matrices are 76, 216, 340, and 540 for Haar, CDF(2,2), CDF(2,4), and Battle-Lemarie based systems, respectively.

4.2   2D APPLICATION: PROPAGATION CHARACTERISTICS OF GENERAL GUIDED WAVE STRUCTURES

The determination of the dispersion characteristics of waveguiding structures is a fundamental problem in microwave engineering applications. Furthermore, these characteristics (mode patterns, propagation constant, characteristic impedance, etc.) can be used as port data for 3D simulation of microwave devices [38].

Several efforts were made by researchers to solve the problem by using three-dimensional finite difference techniques [39, 40]. However, these techniques consume too much computational power and can only calculate single mode characteristics and patterns. Later efforts were able to improve these techniques into two-dimensional formulations called the Compact 2D FDTD method [41, 42, 43] that later led to methods that were able to solve multimode patterns [44, 45]. Despite having computational efficiency, the compact 2D FDTD method requires propagation constant β as an input and depends on signal processing techniques to calculate multimode parameters. It is also prone to stability problems [46].

A recently introduced [47] and further improved [48, 49, 50, 51] compact 2D FDFD technique is more suited for the problem at hand, since it does not require the propagation constant as an input, calculates multimode characteristics without signal processing techniques, and is immune to stability considerations. In this section, characterization of guided wave structures is considered for validation of the proposed MRFD method.

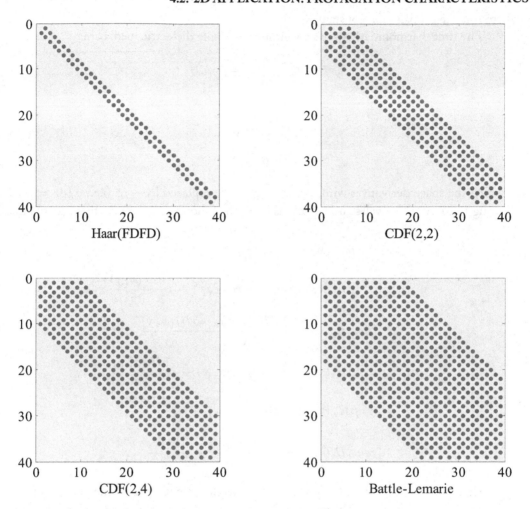

Figure 4.5: Sparsity pattern of coefficient matrices for various MRFD schemes.

FORMULATION

Capitalizing on the compact 2D FDFD method [47], the compact 2D MRFD formulation is derived. Assuming that the waveguiding structure is uniform along the z axis and the wave is propagating in the positive z direction, the electric and magnetic fields inside guided wave structure can be expressed as:

$$\vec{E}(x, y, z) = \left[E_x(x, y)\hat{x} + E_y(x, y)\hat{y} + E_z(x, y)\hat{z} \right] e^{-j\beta z} \tag{4.6a}$$

$$\vec{H}(x, y, z) = \left[H_x(x, y)\hat{x} + H_y(x, y)\hat{y} + H_z(x, y)\hat{z} \right] e^{-j\beta z} \tag{4.6b}$$

where β is the propagation constant.

The time-harmonic Maxwell's equations for simple dielectric media are:

$$\nabla \times \vec{E} = -j\omega\mu\,\vec{H} \tag{4.7a}$$

$$\nabla \times \vec{H} = j\omega\varepsilon\,\vec{E} \tag{4.7b}$$

$$\nabla \cdot \vec{D} = 0 \tag{4.7c}$$

$$\nabla \cdot \vec{B} = 0 \tag{4.7d}$$

in which the space derivatives with respect to z can be replaced by $-j\beta$ (i.e., $\partial/\partial z = -j\beta$). Substituting (4.6) into Maxwell's curl equations (4.7), the following scalar equations are obtained:

$$\beta E_x(x, y) = \omega\mu_y H_y(x, y) + j\frac{\partial E_z(x, y)}{\partial x} \tag{4.8a}$$

$$\beta E_y(x, y) = -\omega\mu_x H_x(x, y) + j\frac{\partial E_z(x, y)}{\partial y} \tag{4.8b}$$

$$j\omega\varepsilon_z E_z(x, y) = \frac{\partial H_y(x, y)}{\partial x} - \frac{\partial H_x(x, y)}{\partial y} \tag{4.8c}$$

$$\beta H_x(x, y) = -\omega\varepsilon_y E_y(x, y) + j\frac{\partial H_z(x, y)}{\partial x} \tag{4.8d}$$

$$\beta H_y(x, y) = \omega\varepsilon_x E_x(x, y) + j\frac{\partial H_z(x, y)}{\partial y} \tag{4.8e}$$

$$j\omega\mu_z H_z(x, y) = -\frac{\partial E_y(x, y)}{\partial x} + \frac{\partial E_x(x, y)}{\partial y}. \tag{4.8f}$$

Equations (4.8c) and (4.8f) don't include β, so they are useless for the purpose of calculating β. The remaining four equations are insufficient to solve the six unknown fields. The needed two additional equations can be supplied by substituting (4.6) into Maxwell's divergence equations, which in return yields:

$$\beta\varepsilon_z E_z(x, y) = -j\varepsilon_x\frac{\partial E_x(x, y)}{\partial x} - j\varepsilon_y\frac{\partial E_y(x, y)}{\partial y} \tag{4.9a}$$

$$\beta\mu_z H_z(x, y) = -j\mu_x\frac{\partial H_x(x, y)}{\partial x} - j\mu_y\frac{\partial H_y(x, y)}{\partial y}. \tag{4.9b}$$

The obtained six scalar equations (4.8a), (4.8b), (4.8d), (4.8e), (4.9a), and (4.9b) are sufficient to solve the problem at hand.

The discretization of these equations is not feasible by simplifying the 3D-MRFD update equations (3.28) previously introduced, because new terms such as $\partial E_x/\partial x$ are not provided by these equations. It is necessary to derive the compact 2D-MRFD formulation. MRFD formulations for equations (4.8a) and (4.9a) are derived here for the sake of illustration.

The derivation again starts with the expansion of the unknown fields in terms of the dual scaling functions, such that:

$$E_x(x, y) = \sum_{i', j'} E_x(i', j') \tilde{\phi}_{i'+1/2}(x) \tilde{\phi}_{j'}(y) \tag{4.10a}$$

$$E_y(x, y) = \sum_{i', j'} E_y(i', j') \tilde{\phi}_{i'}(x) \tilde{\phi}_{j'+1/2}(y) \tag{4.10b}$$

$$E_z(x, y) = \sum_{i', j'} E_z(i', j') \tilde{\phi}_{i'}(x) \tilde{\phi}_{j'}(y) \tag{4.10c}$$

$$H_x(x, y) = \sum_{i', j'} H_x(i', j') \tilde{\phi}_{i'}(x) \tilde{\phi}_{j'+1/2}(y) \tag{4.10d}$$

$$H_y(x, y) = \sum_{i', j'} H_y(i', j') \tilde{\phi}_{i'+1/2}(x) \tilde{\phi}_{j'}(y) \tag{4.10e}$$

$$H_z(x, y) = \sum_{i', j'} H_z(i', j') \tilde{\phi}_{i'+1/2}(x) \tilde{\phi}_{j'+1/2}(y) \; . \tag{4.10f}$$

The locations of the scaling functions in the above expansions correspond to the locations of the field components on the compact 2D grid shown in Figure 4.6. This grid structure is obtained simply by collapsing the 3D Yee cell in the z direction.

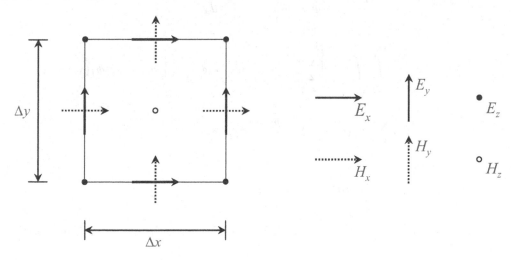

Figure 4.6: 2D compact unit cell.

The sampling procedure starts with testing the left-hand side of Eq. (4.8a) with $\phi_{i+1/2}(x)\phi_j(y)$:

$$\int_{-\infty}^{\infty}\int_{-\infty}^{\infty} \beta E_x(x,y)\phi_{i+1/2}(x)\phi_j(y)dxdy$$

$$= \beta \int_{-\infty}^{\infty}\int_{-\infty}^{\infty} \sum_{i',j'} E_x(i',j')\tilde{\phi}_{i'+1/2}(x)\tilde{\phi}_{j'}(y)\phi_{i+1/2}(x)\phi_j(y)dxdy \qquad (4.11)$$

$$= \beta E_x(i,j)\Delta x \Delta y.$$

Next, the first term on the right-hand side of (4.8a) is tested:

$$\int_{-\infty}^{\infty}\int_{-\infty}^{\infty} \omega\mu_y H_y(x,y)\phi_{i+1/2}(x)\phi_j(y)dxdy$$

$$= \omega\mu_y \int_{-\infty}^{\infty}\int_{-\infty}^{\infty} \sum_{i',j'} H_y(i',j')\tilde{\phi}_{i'+1/2}(x)\tilde{\phi}_{j'}(y)\phi_{i+1/2}(x)\phi_j(y)dxdy \qquad (4.12)$$

$$= \omega\mu_y H_y(i,j)\Delta x \Delta y,$$

and finally, the second-term on the right-hand side of (4.8a) is sampled to yield:

$$\int_{-\infty}^{\infty}\int_{-\infty}^{\infty} j\frac{\partial E_z(x,y)}{\partial x}\phi_{i+1/2}(x)\phi_j(y)dxdy$$

$$= j \int_{-\infty}^{\infty}\int_{-\infty}^{\infty} \frac{\partial}{\partial x}\left[\sum_{i',j'} E_z(i',j')\tilde{\phi}_{i'}(x)\tilde{\phi}_{j'}(y)\right]\phi_{i+1/2}(x)\phi_j(y)dxdy$$

$$= j\sum_{i',j'} E_z(i',j') \int_{-\infty}^{\infty}\frac{\partial\tilde{\phi}_{i'}(x)}{\partial x}\phi_{i+1/2}(x)dx \int_{-\infty}^{\infty}\tilde{\phi}_{j'}(y)\phi_j(y)dy$$

$$= j\Delta y \sum_{l=1}^{3} a(l)\left[E_z(i+l,j) - E_z(i-l+1,j)\right]. \qquad (4.13)$$

Substituting (4.11), (4.12), and (4.13) into (4.8a) yields the MRFD update equation:

$$\beta E_x(i,j) = \omega\mu H_y(i,j) + j\frac{\sum_{l=1}^{3} a(l)\left[E_z(i+l,j) - E_z(i-l+1,j)\right]}{\Delta x}. \qquad (4.14)$$

The update equation for (4.9a) is similarly derived, first by testing the left-hand side:

$$\int_{-\infty}^{\infty}\int_{-\infty}^{\infty} \beta\varepsilon_z E_z(x,y)\phi_i(x)\phi_j(y)dxdy$$

$$= \int_{-\infty}^{\infty}\int_{-\infty}^{\infty} \beta\varepsilon_z\left[\sum_{i',j'} E_z(i',j')\tilde{\phi}_{i'}(x)\tilde{\phi}_{j'}(y)\right]\phi_i(x)\phi_j(y)dxdy \qquad (4.15)$$

$$= \beta\varepsilon_z E_z(i,j)\Delta x\Delta y.$$

Testing the first term on the right-hand side of (4.9a) yields:

$$\int_{-\infty}^{\infty}\int_{-\infty}^{\infty} -j\varepsilon_x \frac{\partial E_x(x,y)}{\partial x}\phi_i(x)\phi_j(y)dxdy$$

$$= -j\varepsilon_x\int_{-\infty}^{\infty}\int_{-\infty}^{\infty}\frac{\partial}{\partial x}\left[\sum_{i',j'} E_x(i',j')\tilde{\phi}_{i'+1/2}(x)\tilde{\phi}_{j'}(y)\right]\phi_i(x)\phi_j(y)dxdy \qquad (4.16)$$

$$= -j\varepsilon_x\sum_{i',j'} E_x(i',j')\int_{-\infty}^{\infty}\frac{\partial\tilde{\phi}_{i'+1/2}(x)}{\partial x}\phi_i(x)dx\int_{-\infty}^{\infty}\tilde{\phi}_{j'}(y)\phi_j(y)dy$$

$$= -j\varepsilon_x\Delta y\sum_{l=1}^{3} a(l)\left[E_x(i+l-1,j) - E_x(i-l,j)\right].$$

Similarly, the second term on the right-hand side of (4.9a) is sampled to give:

$$\int_{-\infty}^{\infty}\int_{-\infty}^{\infty} -j\varepsilon_y\frac{\partial E_y(x,y)}{\partial y}\phi_i(x)\phi_j(y)dxdy$$

$$= -j\varepsilon_y\sum_{l=1}^{3} a(l)\left[E_y(i,j+l-1) - E_y(i,j-l)\right]. \qquad (4.17)$$

Finally, substituting (4.15), (4.16), and (4.17) into (4.9a) yields the MRFD update equation:

$$\beta\varepsilon_z E_z(i,j) = -j\frac{\displaystyle\sum_{l=1}^{3} a(l)\varepsilon_x\left[E_x(i+l-1,j) - E_x(i-l,j)\right]}{\Delta x}$$

$$-j\frac{\displaystyle\sum_{l=1}^{3} a(l)\varepsilon_y\left[E_y(i,j+l-1) - E_y(i,j-l)\right]}{\Delta y}. \qquad (4.18)$$

The remaining update equations can be generated similarly. The complete set of equations of the compact 2D MRFD formulation is listed below:

$$\beta E_x(i, j) = \omega \mu_y H_y(i, j) + j \frac{\sum\limits_{l=1}^{3} a(l) \left[E_z(i+l, j) - E_z(i-l+1, j) \right]}{\Delta x} \tag{4.19a}$$

$$\beta E_y(i, j) = - \omega \mu_x H_x(i, j) + j \frac{\sum\limits_{l=1}^{3} a(l) \left[E_z(i, j+l) - E_z(i, j-l+1) \right]}{\Delta y} \tag{4.19b}$$

$$\beta \varepsilon_z(i, j) E_z(i, j) = - j \frac{\sum\limits_{l=1}^{3} a(l)\varepsilon_x \left[E_x(i+l-1, j) - E_x(i-l, j) \right]}{\Delta x}$$

$$- j \frac{\sum\limits_{l=1}^{3} a(l)\varepsilon_y \left[E_y(i, j+l-1) - E_y(i, j-l) \right]}{\Delta y} \tag{4.19c}$$

$$\beta H_x(i, j) = - \omega \varepsilon_y E_y(i, j) + j \frac{\sum\limits_{l=1}^{3} a(l) \left[H_z(i+l-1, j) - H_z(i-l, j) \right]}{\Delta x} \tag{4.19d}$$

$$\beta H_y(i, j) = \omega \varepsilon_x E_x(i, j) + j \frac{\sum\limits_{l=1}^{3} a(l) \left[H_z(i, j+l-1) - H_z(i, j-l) \right]}{\Delta y} \tag{4.19e}$$

$$\beta \mu_z(i, j) H_z(i, j) = - j \frac{\sum\limits_{l=1}^{3} a(l)\mu_x \left[H_x(i+l, j) - H_x(i-l+1, j) \right]}{\Delta x}$$

$$- j \frac{\sum\limits_{l=1}^{3} a(l)\mu_y \left[H_y(i, j+l) - H_y(i, j-l+1) \right]}{\Delta y}. \tag{4.19f}$$

The material properties can be sampled pointwise similar to the previous cases. After assigning the material properties and arranging the terms, the finalized compact 2D MRFD formulation becomes:

$$\beta E_x(i, j) = \omega \mu_y(i, j) H_y(i, j) + c_{1x} E_z(i+1, j) - c_{1x} E_z(i, j)$$

$$+ c_{2x} E_z(i+2, j) - c_{2x} E_z(i-1, j)$$

$$+ c_{3x} E_z(i+3, j) - c_{3x} E_z(i-2, j) \tag{4.20a}$$

$$\beta E_y(i, j) = -\omega\mu_x(i, j)H_x(i, j) + c_{1y}E_z(i, j + 1) - c_{1y}E_z(i, j)$$
$$+ c_{2y}E_z(i, j + 2) - c_{2y}E_z(i, j - 1)$$
$$+ c_{3y}E_z(i, j + 3) - c_{3y}E_z(i, j - 2) \tag{4.20b}$$

$$\beta E_z(i, j) = -c_{1x}\frac{\varepsilon_x(i, j)}{\varepsilon_z(i, j)}E_x(i, j) + c_{1x}\frac{\varepsilon_x(i - 1, j)}{\varepsilon_z(i, j)}E_x(i - 1, j)$$
$$- c_{2x}\frac{\varepsilon_x(i + 1, j)}{\varepsilon_z(i, j)}E_x(i + 1, j) + c_{2x}\frac{\varepsilon_x(i - 2, j)}{\varepsilon_z(i, j)}E_x(i - 2, j)$$
$$- c_{3x}\frac{\varepsilon_x(i + 2, j)}{\varepsilon_z(i, j)}E_x(i + 2, j) + c_{3x}\frac{\varepsilon_x(i - 3, j)}{\varepsilon_z(i, j)}E_x(i - 3, j)$$
$$- c_{1y}\frac{\varepsilon_y(i, j)}{\varepsilon_z(i, j)}E_y(i, j) + c_{1y}\frac{\varepsilon_y(i, j - 1)}{\varepsilon_z(i, j)}E_y(i, j - 1)$$
$$- c_{2y}\frac{\varepsilon_y(i, j + 1)}{\varepsilon_z(i, j)}E_y(i, j + 1) + c_{2y}\frac{\varepsilon_y(i, j - 2)}{\varepsilon_z(i, j)}E_y(i, j - 2)$$
$$- c_{3y}\frac{\varepsilon_y(i, j + 2)}{\varepsilon_z(i, j)}E_y(i, j + 2) + c_{3y}\frac{\varepsilon_y(i, j - 3)}{\varepsilon_z(i, j)}E_y(i, j - 3) \tag{4.20c}$$

$$\beta H_x(i, j) = -\omega\varepsilon_y(i, j)E_y(i, j) + c_{1x}H_z(i, j) - c_{1x}H_z(i - 1, j)$$
$$+ c_{2x}H_z(i + 1, j) - c_{2x}H_z(i - 2, j)$$
$$+ c_{3x}H_z(i + 2, j) - c_{3x}H_z(i - 3, j) \tag{4.20d}$$

$$\beta H_y(i, j) = \omega\varepsilon_x(i, j)E_x(i, j) + c_{1y}H_z(i, j) - c_{1y}H_z(i, j - 1)$$
$$+ c_{2y}H_z(i, j + 1) - c_{2y}H_z(i, j - 2)$$
$$+ c_{3y}H_z(i, j + 2) - c_{3y}H_z(i, j - 3) \tag{4.20e}$$

$$\beta H_z(i, j) = -c_{1x}\frac{\mu_x(i + 1, j)}{\mu_z(i, j)}H_x(i + 1, j) + c_{1x}\frac{\mu_x(i, j)}{\mu_z(i, j)}H_x(i, j)$$
$$- c_{2x}\frac{\mu_x(i + 2, j)}{\mu_z(i, j)}H_x(i + 2, j) + c_{2x}\frac{\mu_x(i - 1, j)}{\mu_z(i, j)}H_x(i - 1, j)$$
$$- c_{3x}\frac{\mu_x(i + 3, j)}{\mu_z(i, j)}H_x(i + 3, j) + c_{3x}\frac{\mu_x(i - 2, j)}{\mu_z(i, j)}H_x(i - 2, j)$$
$$- c_{1y}\frac{\mu_y(i, j + 1)}{\mu_z(i, j)}H_y(i, j + 1) + c_{1y}\frac{\mu_y(i, j)}{\mu_z(i, j)}H_y(i, j)$$

$$-c_{2y}\frac{\mu_y(i, j+2)}{\mu_z(i, j)}H_y(i, j+2) + c_{2y}\frac{\mu_y(i, j-1)}{\mu_z(i, j)}H_y(i, j-1)$$

$$-c_{3y}\frac{\mu_y(i, j+3)}{\mu_z(i, j)}H_y(i, j+3) + c_{3y}\frac{\mu_y(i, j-2)}{\mu_z(i, j)}H_y(i, j-2). \tag{4.20f}$$

The update coefficients listed below are introduced in order to simplify the above formulation.

$$
\begin{aligned}
c_{1x} &= ja(1)/\Delta x, & c_{1y} &= ja(1)/\Delta y, \\
c_{2x} &= ja(2)/\Delta x, & c_{2y} &= ja(2)/\Delta y, \\
c_{3x} &= ja(3)/\Delta x, & c_{3y} &= ja(3)/\Delta y.
\end{aligned}
\tag{4.21}
$$

The compact 2D MRFD update equations given by (4.20) can be used to form an eigenvalue problem:

$$\beta.x = A.x. \tag{4.22}$$

Here, A is the sparse coefficient matrix and x is the unknown field vector. The eigenvalues of A deliver the propagation constant and the eigenvectors of A deliver the corresponding mode patterns.

THE MULTIPLE IMAGE TECHNIQUE

In Section 4.1, the treatment of PEC boundary conditions in the context of image principle was successfully utilized. Image principle was used to extend the electromagnetic fields beyond the computational domain by forcing odd symmetry for tangential electric fields and normal magnetic fields, whereas even symmetry was forced for normal electric fields and tangential magnetic fields.

The material parameters are part of the update equations, and thus they should also be extended beyond the boundary. The concept of multiple image technique (MIT) was developed in order to handle the extension of material parameters [52]. This technique assumes a mirror image of the materials along the PEC boundary. For a boxed microstrip line, the multiple image technique can be applied as depicted in Figure 4.7. In this figure, the original problem to be analyzed is at the center, whereas the mirrored images are generated on the top, bottom, and sides of the original problem. Note that this technique does not increase memory consumption.

The field symmetries that are forced on the boundary of the microstrip line are listed in Table 4.3.

NUMERICAL RESULTS

Several waveguiding structures are analyzed by both FDFD and MRFD methods. The first example is a dielectric filled waveguide shown in Figure 4.8 with dimensions $a = 0.6$ cm and $b = 1.5$ cm. The propagation constants of the TE_{01} and TE_{02} modes of this waveguide structure are analyzed

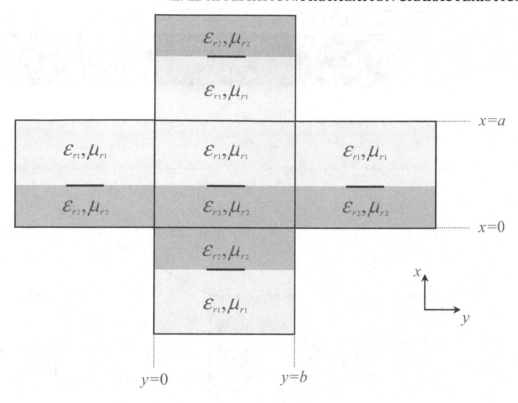

Figure 4.7: Cross section of the boxed microstrip line with its images.

with the aid of FDFD and MRFD methods. Obtained numerical results are compared to analytical values [53] in Figure 4.9. Computation of the exact propagation constants of TE_{01} and TE_{02} modes of this waveguide structure is summarized in Appendix B.

For the MRFD simulation of the waveguide, a cell size of 0.3 cm, which is three times larger than the FDFD cell size, is chosen, reducing the total number of grid points by a factor of 9. Accuracy of both methods is found to be similar for the first propagation mode. For the second mode, the MRFD scheme returned slightly more accurate results.

The computer resources consumed by both methods are summarized in Table 4.4, which shows that the MRFD method improves on the FDFD by 86.44% in terms of memory efficiency and 82.70% in terms of simulation time.

The second example is a partially dielectric loaded waveguide shown in Figure 4.10. The physical dimensions of this structure are $a = 0.6\,\text{cm}, b = 1.5\,\text{cm}$, and $h = 0.3$ cm. The propagation constants of the TM_{X01} and TM_{X02} modes of the structure are calculated by FDFD and MRFD methods. Obtained numerical results are compared to analytical values [53] in Figure 4.11. Compu-

Table 4.3: Fields symmetries across the PEC boundaries

	Symmetry Plane			
	$x=0$	$x=a$	$y=0$	$y=b$
E_x, H_y	even	even	odd	odd
E_y, H_x	odd	odd	even	even
E_z	odd	odd	odd	odd
H_z	even	even	even	even

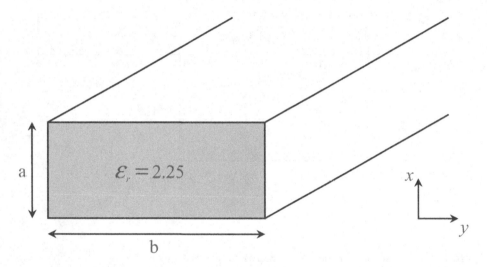

Figure 4.8: Dielectric filled rectangular waveguide.

tation of the exact propagation constants of TM_{X01} and TM_{X02} modes of this waveguide structure is summarized in Appendix B.

Similar to the homogeneously filled waveguide case, accuracy of both methods is found to be similar for the first propagation mode, whereas the MRFD scheme returned slightly more accurate results for the second propagation mode.

The efficiency of both methods is compared in Table 4.5. In this case, the MRFD method provides improvements of 64.13% in terms of memory efficiency and 62.87% in terms of simulation time, over the FDFD method.

The last example considered is a boxed microstrip line depicted in Figure 4.12. The physical dimensions of this structure are $a = 0.6$ cm, $b = 1.5$ cm, $h = 0.3$ cm, and $w = 0.3$ cm. The propagation constant and effective dielectric constant of the dominant mode of this structure is calculated by FDFD and MRFD methods together with CST Microwave Studio, a commercial 3D solver.

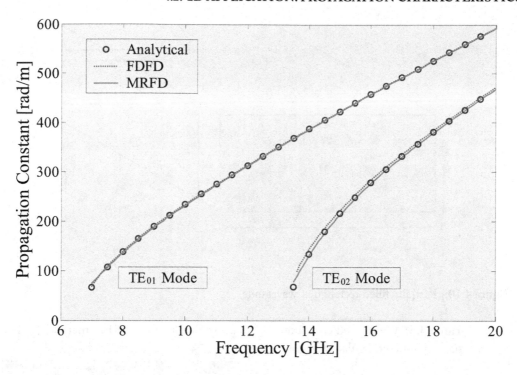

Figure 4.9: Propagation constant of the dielectrice filled rectangular waveguide.

Table 4.4: Consumed computer resources by FDFD and MRFD methods, dielectric filled rectangular waveguide case

	FDFD	MRFD
Number of Cells	15x6	5x2
Matrix Size (bytes)	32336	4384
Simulation Time per Freq. (sec)	1.549	0.268
Memory Savings	-	86.44%
Time Savings	-	82.70%

The calculated propagation constant and effective dielectric constant are presented in Figure 4.13 and Figure 4.14, respectively. Accuracy of both methods is found to be similar, except at the lower frequency where the MRFD method is found to be more accurate.

The efficiency figures of both methods are shown in Table 4.6. In this case, the MRFD method improved on the FDFD method by 17.68% in terms of memory efficiency and 35.04% in terms of simulation time.

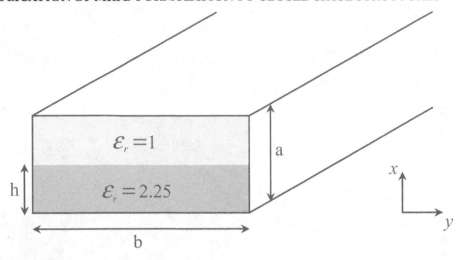

Figure 4.10: Partially filled rectangular waveguide.

	FDFD	MRFD
Table 4.5: Consumed computer resources by FDFD and MRFD methods, partially filled rectangular waveguide case		
Number of Cells	15x6	5x4
Matrix Size (bytes)	32336	11600
Simulation Time per Freq. (sec)	2.001	0.743
Memory Savings	-	64.13%
Time Savings	-	62.87%

It appears that the increasing complexity of the problem has a negative effect on the efficiency of the MRFD technique. In reality, as the complexity increases, we have chosen finer grids to match the FDFD results. The complexity of the problem has no relative effect on matrix size or computation time needed to solve the matrix.

4.3   3D APPLICATION: THE RECTANGULAR CAVITY RESONATOR

The rectangular cavity resonator structure is the choice of the three-dimensional example, again because this kind of structure is best suited to a frequency domain application. In this section, homogeneously and inhomogeneously filled cavity resonators are simulated by both MRFD and FDFD methods.

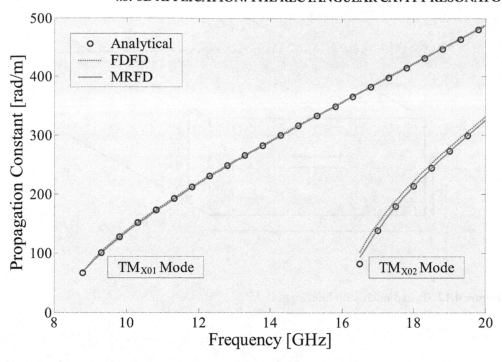

Figure 4.11: Propagation constant of the partially filled rectangular waveguide.

	FDFD	MRFD
Table 4.6: Consumed computer resources by FDFD and MRFD methods, boxed microstrip line		
Number of Cells	15x6	10x4
Matrix Size (bytes)	32336	26620
Simulation Time per Freq. (sec)	1.424	0.925
Memory Savings	-	17.68%
Time Savings	-	35.04%

FORMULATION

The 3D general MRFD update equations (3.28) can be directly used to generate an eigenvalue problem. After the expansion of summations, the MRFD update equations can be expressed as:

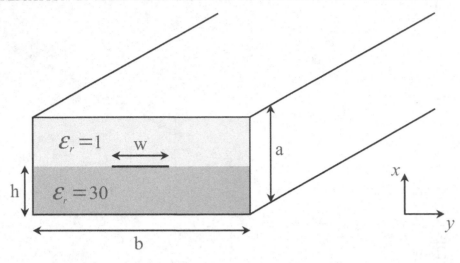

Figure 4.12: Boxed microstrip line.

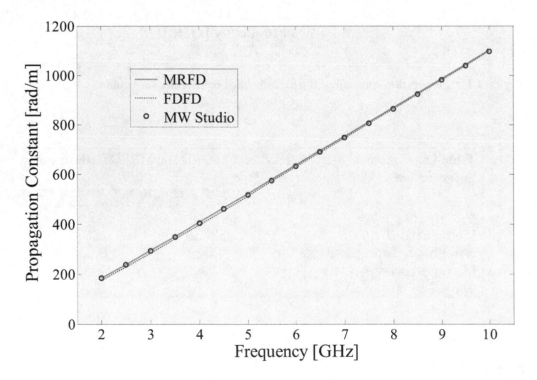

Figure 4.13: Propagation constant of the boxed microstrip line.

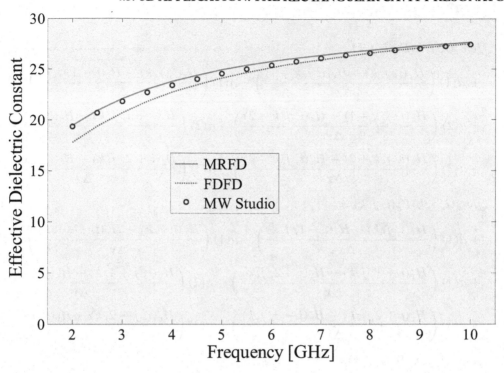

Figure 4.14: Effective dielectric constant of the boxed microstrip line.

$$j\omega\varepsilon_x(i, j, k)E_x(i, j, k) = \tag{4.23a}$$

$$+ a(1)\left(\frac{H_z(i, j, k) - H_z(i, j - 1, k)}{\Delta y}\right) - a(1)\left(\frac{H_y(i, j, k) - H_y(i, j, k - 1)}{\Delta z}\right)$$

$$+ a(2)\left(\frac{H_z(i, j + 1, k) - H_z(i, j - 2, k)}{\Delta y}\right) - a(2)\left(\frac{H_y(i, j, k + 1) - H_y(i, j, k - 2)}{\Delta z}\right)$$

$$+ a(3)\left(\frac{H_z(i, j + 2, k) - H_z(i, j - 3, k)}{\Delta y}\right) - a(3)\left(\frac{H_y(i, j, k + 2) - H_y(i, j, k - 3)}{\Delta z}\right)$$

$$j\omega\varepsilon_y(i,j,k)E_y(i,j,k) = \tag{4.23b}$$

$$+ a(1)\left(\frac{H_x(i,j,k) - H_x(i,j,k-1)}{\Delta z}\right) - a(1)\left(\frac{H_z(i,j,k) - H_z(i-1,j,k)}{\Delta x}\right)$$

$$+ a(2)\left(\frac{H_x(i,j,k+1) - H_x(i,j,k-2)}{\Delta z}\right) - a(2)\left(\frac{H_z(i+1,j,k) - H_z(i-2,j,k)}{\Delta x}\right)$$

$$+ a(3)\left(\frac{H_x(i,j,k+2) - H_x(i,j,k-3)}{\Delta z}\right) - a(3)\left(\frac{H_z(i+2,j,k) - H_z(i-3,j,k)}{\Delta x}\right)$$

$$j\omega\varepsilon_z(i,j,k)E_z(i,j,k) = \tag{4.23c}$$

$$+ a(1)\left(\frac{H_y(i,j,k) - H_y(i-1,j,k)}{\Delta x}\right) - a(1)\left(\frac{H_x(i,j,k) - H_x(i,j-1,k)}{\Delta y}\right)$$

$$+ a(2)\left(\frac{H_y(i+1,j,k) - H_y(i-2,j,k)}{\Delta x}\right) - a(2)\left(\frac{H_x(i,j+1,k) - H_x(i,j-2,k)}{\Delta y}\right)$$

$$+ a(3)\left(\frac{H_y(i+2,j,k) - H_y(i-3,j,k)}{\Delta x}\right) - a(3)\left(\frac{H_x(i,j+2,k) - H_x(i,j-3,k)}{\Delta y}\right)$$

$$j\omega\mu_x(i,j,k)H_x(i,j,k) = \tag{4.23d}$$

$$+ a(1)\left(\frac{E_y(i,j,k+1) - E_y(i,j,k)}{\Delta z}\right) - a(1)\left(\frac{E_z(i,j+1,k) - E_z(i,j,k)}{\Delta y}\right)$$

$$+ a(2)\left(\frac{E_y(i,j,k+2) - E_y(i,j,k-1)}{\Delta z}\right) - a(2)\left(\frac{E_z(i,j+2,k) - E_z(i,j-1,k)}{\Delta y}\right)$$

$$+ a(3)\left(\frac{E_y(i,j,k+3) - E_y(i,j,k-2)}{\Delta z}\right) - a(3)\left(\frac{E_z(i,j+3,k) - E_z(i,j-2,k)}{\Delta y}\right)$$

$$j\omega\mu_y(i,j,k)H_y(i,j,k) = \tag{4.23e}$$

$$+ a(1)\left(\frac{E_z(i+1,j,k) - E_z(i,j,k)}{\Delta x}\right) - a(1)\left(\frac{E_x(i,j,k+1) - E_x(i,j,k)}{\Delta z}\right)$$

$$+ a(2)\left(\frac{E_z(i+2,j,k) - E_z(i-1,j,k)}{\Delta x}\right) - a(2)\left(\frac{E_x(i,j,k+2) - E_x(i,j,k-1)}{\Delta z}\right)$$

$$+ a(3)\left(\frac{E_z(i+3,j,k) - E_z(i-2,j,k)}{\Delta x}\right) - a(3)\left(\frac{E_x(i,j,k+3) - E_x(i,j,k-2)}{\Delta z}\right)$$

$$j\omega\mu_z(i, j, k)H_z(i, j, k) = \qquad (4.23\mathrm{f})$$

$$+ a(1)\left(\frac{E_x(i, j+1, k) - E_x(i, j, k)}{\Delta y}\right) - a(1)\left(\frac{E_y(i+1, j, k) - E_y(i, j, k)}{\Delta x}\right)$$

$$+ a(2)\left(\frac{E_x(i, j+2, k) - E_x(i, j-1, k)}{\Delta y}\right) - a(2)\left(\frac{E_y(i+2, j, k) - E_y(i-1, j, k)}{\Delta x}\right)$$

$$+ a(3)\left(\frac{E_x(i, j+3, k) - E_x(i, j-2, k)}{\Delta y}\right) - a(3)\left(\frac{E_y(i+3, j, k) - E_y(i-2, j, k)}{\Delta x}\right).$$

These equations can be organized into an eigenvalue problem as follows:

$$\omega E_x(i, j, k) = \qquad (4.24\mathrm{a})$$

$$- C_{exy1}(m)\left[H_z(i, j, k) - H_z(i, j-1, k)\right] + C_{exz1}(m)\left[H_y(i, j, k) - H_y(i, j, k-1)\right]$$

$$- C_{exy2}(m)\left[H_z(i, j+1, k) - H_z(i, j-2, k)\right] + C_{exz2}(m)\left[H_y(i, j, k+1) - H_y(i, j, k-2)\right]$$

$$- C_{exy3}(m)\left[H_z(i, j+2, k) - H_z(i, j-3, k)\right] + C_{exz3}(m)\left[H_y(i, j, k+2) - H_y(i, j, k-3)\right]$$

$$\omega E_y(i, j, k) = \qquad (4.24\mathrm{b})$$

$$- C_{eyz1}(m)\left[H_x(i, j, k) - H_x(i, j, k-1)\right] + C_{eyx1}(m)\left[H_z(i, j, k) - H_z(i-1, j, k)\right]$$

$$- C_{eyz2}(m)\left[H_x(i, j, k+1) - H_x(i, j, k-2)\right] + C_{eyx2}(m)\left[H_z(i+1, j, k) - H_z(i-2, j, k)\right]$$

$$- C_{eyz3}(m)\left[H_x(i, j, k+2) - H_x(i, j, k-3)\right] + C_{eyx3}(m)\left[H_z(i+2, j, k) - H_z(i-3, j, k)\right]$$

$$\omega E_z(i, j, k) = \qquad (4.24\mathrm{c})$$

$$- C_{ezx1}(m)\left[H_y(i, j, k) - H_y(i-1, j, k)\right] + C_{ezy1}(m)\left[H_x(i, j, k) - H_x(i, j-1, k)\right]$$

$$- C_{ezx2}(m)\left[H_y(i+1, j, k) - H_y(i-2, j, k)\right] + C_{ezy2}(m)\left[H_x(i, j+1, k) - H_x(i, j-2, k)\right]$$

$$- C_{ezx3}(m)\left[H_y(i+2, j, k) - H_y(i-3, j, k)\right] + C_{ezy3}(m)\left[H_x(i, j+2, k) - H_x(i, j-3, k)\right]$$

$$\omega H_x(i, j, k) = \qquad (4.24\mathrm{d})$$

$$- C_{hxz1}(m)\left[E_y(i, j, k+1) - E_y(i, j, k)\right] + C_{hxy1}(m)\left[E_z(i, j+1, k) - E_z(i, j, k)\right]$$

$$- C_{hxz2}(m)\left[E_y(i, j, k+2) - E_y(i, j, k-1)\right] + C_{hxy2}(m)\left[E_z(i, j+2, k) - E_z(i, j-1, k)\right]$$

$$- C_{hxz3}(m)\left[E_y(i, j, k+3) - E_y(i, j, k-2)\right] + C_{hxy3}(m)\left[E_z(i, j+3, k) - E_z(i, j-2, k)\right]$$

$$\omega H_y(i, j, k) = \qquad (4.24e)$$

$$- C_{hyx1}(m)\left[E_z(i+1, j, k) - E_z(i, j, k)\right] + C_{hyz1}(m)\left[E_x(i, j, k+1) - E_x(i, j, k)\right]$$

$$- C_{hyx2}(m)\left[E_z(i+2, j, k) - E_z(i-1, j, k)\right] + C_{hyz2}(m)\left[E_x(i, j, k+2) - E_x(i, j, k-1)\right]$$

$$- C_{hyx3}(m)\left[E_z(i+3, j, k) - E_z(i-2, j, k)\right] + C_{hyz3}(m)\left[E_x(i, j, k+3) - E_x(i, j, k-2)\right]$$

$$\omega H_z(i, j, k) = \qquad (4.24f)$$

$$- C_{hzy1}(m)\left[E_x(i, j+1, k) - E_x(i, j, k)\right] + C_{hzx1}(m)\left[E_y(i+1, j, k) - E_y(i, j, k)\right]$$

$$- C_{hzy2}(m)\left[E_x(i, j+2, k) - E_x(i, j-1, k)\right] + C_{hzx2}(m)\left[E_y(i+2, j, k) - E_y(i-1, j, k)\right]$$

$$- C_{hzy3}(m)\left[E_x(i, j+3, k) - E_x(i, j-2, k)\right] + C_{hzx3}(m)\left[E_y(i+3, j, k) - E_y(i-2, j, k)\right]$$

with the field update coefficients of:

$$
\begin{aligned}
C_{exyi}(m) &= \frac{ja(i)}{\varepsilon_x(m)\Delta y} & C_{exzi}(m) &= \frac{ja(i)}{\varepsilon_x(m)\Delta z} \\
C_{eyzi}(m) &= \frac{ja(i)}{\varepsilon_y(m)\Delta z} & C_{eyxi}(m) &= \frac{ja(i)}{\varepsilon_y(m)\Delta x} \\
C_{ezxi}(m) &= \frac{ja(i)}{\varepsilon_z(m)\Delta x} & C_{ezyi}(m) &= \frac{ja(i)}{\varepsilon_z(m)\Delta y} \\
C_{hxyi}(m) &= \frac{ja(i)}{\mu_x(m)\Delta y} & C_{hxzi}(m) &= \frac{ja(i)}{\mu_x(m)\Delta z} \\
C_{hyxi}(m) &= \frac{ja(i)}{\mu_y(m)\Delta x} & C_{hyzi}(m) &= \frac{ja(i)}{\mu_y(m)\Delta z} \\
C_{hzxi}(m) &= \frac{ja(i)}{\mu_z(m)\Delta x} & C_{hzyi}(m) &= \frac{ja(i)}{\mu_z(m)\Delta y}
\end{aligned}
\qquad (4.25)
$$

where $m = (i, j, k)$.

The MRFD update equations given by (4.24) can be used to form an eigenvalue problem, such as:

$$\omega.x = A.x. \qquad (4.26)$$

Again, A is the sparse coefficient matrix and x is the unknown field vector. The eigenvalues of A deliver the resonance frequencies, and the eigenvectors of A deliver the corresponding mode patterns.

NUMERICAL RESULTS

The first example considered is an air-filled cavity resonator shown in Figure 4.15. The dimensions of this resonator are selected to be $a = 10$ cm, $b = 15$ cm, and $c = 20$ cm. A cell size of $\Delta x = \Delta y = \Delta z = 2.5$ cm is used for the FDFD discretization, giving a total number of 192 cells. The MRFD scheme on the other hand, used a mesh with the cell size of $\Delta x = \Delta y = \Delta z = 5$ cm which results in a total of 24 cells.

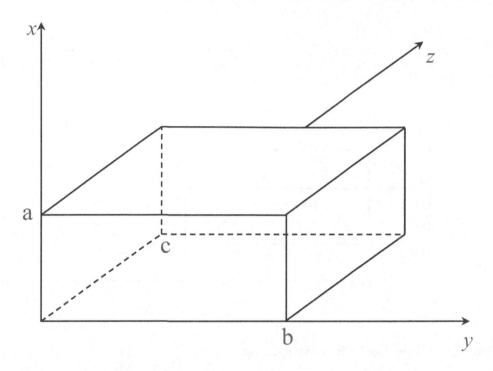

Figure 4.15: Air-filled rectangular cavity resonator.

The resonance frequencies obtained by FDFD and MRFD simulations are compared to the analytical resonance frequencies [53] in Table 4.7. Computation of the exact resonance frequencies of this cavity structure is summarized in Appendix C.

The results are obtained by using a sparse matrix solver and a full matrix solver, both of which are Matlab solvers. The sparse matrix solver is executed by the Matlab *eigs* function and uses the Arnoldi Iteration technique [54]. The full matrix solver is executed by the Matlab *eig* function and uses LAPACK routines [55]. The sparse matrix solver reduced the simulation time and memory requirement by the factors 97.5 and 8, respectively. The full matrix solver, on the other hand, reduced the simulation time by the factor 181.8 and the memory requirement by the factor of 130.9.

The second example considered is a cavity resonator half loaded with dielectric material, depicted in Figure 4.16. Computation of the exact resonance frequencies of this cavity structure is summarized in Appendix C.

The dimensions of the resonator are kept unchanged from the previous example. The cell size for the FDFD mesh also remains unchanged while the cell size for the MRFD mesh in the direction of the discontinuity is set to half of that of the previous case.

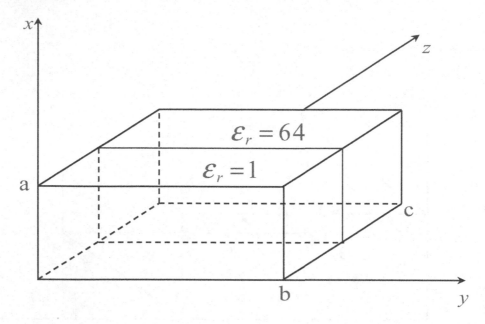

Figure 4.16: Dielectrice loaded cavity resonator.

Table 4.7: Resonance frequencies of the air-filled rectangular cavity				
Analytical Values	FDFD mesh size 4x6x8		MRFD mesh size 2x3x4	
(MHz)	Resonance freq.	Relative error	Resonance freq.	Relative error
1.249	1.237	-0.95%	1.254	0.37%
1.676	1.640	-2.17%	1.694	1.06%
1.801	1.763	-2.11%	1.820	1.04%
1.951	1.914	-1.90%	1.969	0.93%

Numerical results presented in Table 4.8 show that the MRFD scheme provides better accuracy despite using a much smaller mesh. Using a sparse matrix solver, the simulation time and memory requirement are reduced by the factors 52.7 and 3.2, respectively. If a full matrix solver were utilized, the savings changed to the factor 61.7 for simulation time and the factor 27.7 for the memory requirement.

Table 4.8: Resonance frequencies of the rectangular cavity half filled with dielectric material

Analytical Values (MHz)	FDFD mesh size 4x6x8		MRFD mesh size 2x3x8	
	Resonance freq.	Relative error	Resonance freq.	Relative error
186.27	185.18	-0.59%	188.33	1.11%
271.72	267.95	-1.39%	275.98	1.57%
293.75	284.15	-3.27%	297.62	1.32%
350.69	340.37	-2.94%	355.95	1.50%

CHAPTER 5

Application of MRFD Formulation to Open Space Structures

The MRFD technique is used to characterize open space problems in this chapter. A scattered field formulation is developed and implemented for this purpose. The perfectly matched layer (PML) technique is employed in order to terminate the computational space. Numerical results of two-dimensional scattering problems are presented at the end of the chapter.

5.1 GENERAL SCATTERED FIELD FORMULATION

In the previous chapter, the MRFD method is applied to closed space eigenvalue problems where the use of a source field is not necessary. Our objective in this chapter is to investigate the accuracy and efficiency of the MRFD technique that is applicable to the determination of the electromagnetic field distribution in two-dimensional radiation and scattering field problems.

In order to solve scattering and radiation problems, incident electromagnetic waves should be excited in the computation space. The total field/scattered field formulation [13] and the pure scattered field formulation [56] are the two approaches applied to solve these problems. The FDFD and MRFD formulations utilized in this study are based on the scattered field approach. This approach evolves from the linearity of Maxwell's equations and the decomposition of the total $\vec{E}$ and $\vec{H}$ fields into an incident field and a scattered field, that is:

$$\vec{E}_{total} = \vec{E}_{inc} + \vec{E}_{scat} \tag{5.1a}$$

$$\vec{H}_{total} = \vec{H}_{inc} + \vec{H}_{scat}. \tag{5.1b}$$

The incident field is the field that would exist in the computational domain in which no scatterers exist and therefore can be calculated analytically for every cell in the computational domain. The scattered field is the difference between the total field and the incident field in the presence of the scatterer.

If the background of the computational domain is free space, then the incident field satisfies:

$$\nabla \times \vec{E}_{inc} = -j\omega\mu_o \vec{H}_{inc} \tag{5.2a}$$

$$\nabla \times \vec{H}_{inc} = +j\omega\varepsilon_o \vec{E}_{inc} . \tag{5.2b}$$

The total fields satisfy the Maxwell's equations by definition:

$$\nabla \times \vec{E}_{total} = -j\omega\mu \vec{H}_{total} \tag{5.3a}$$

$$\nabla \times \vec{H}_{total} = +j\omega\varepsilon \vec{E}_{total} . \tag{5.3b}$$

Using the scattered field decomposition (5.1), the curl equations can be combined to yield:

$$\nabla \times \vec{E}_{scat} + j\omega\mu \vec{H}_{scat} = j\omega(\mu_o - \mu)\vec{H}_{inc} \tag{5.4a}$$

$$\nabla \times \vec{H}_{scat} - j\omega\varepsilon \vec{E}_{scat} = j\omega(\varepsilon - \varepsilon_o)\vec{E}_{inc} . \tag{5.4b}$$

By decomposing the vector equations into x, y, and z components, the following six scalar equations can be obtained:

$$\frac{\partial E_{scat,z}}{\partial y} - \frac{\partial E_{scat,y}}{\partial z} + j\omega\mu_x H_{scat,x} = j\omega(\mu_o - \mu_x)H_{inc,x} \tag{5.5a}$$

$$\frac{\partial E_{scat,x}}{\partial z} - \frac{\partial E_{scat,z}}{\partial x} + j\omega\mu_y H_{scat,y} = j\omega(\mu_o - \mu_y)H_{inc,y} \tag{5.5b}$$

$$\frac{\partial E_{scat,y}}{\partial x} - \frac{\partial E_{scat,x}}{\partial y} + j\omega\mu_z H_{scat,z} = j\omega(\mu_o - \mu_z)H_{inc,z} \tag{5.5c}$$

$$\frac{\partial H_{scat,z}}{\partial y} - \frac{\partial H_{scat,y}}{\partial z} - j\omega\varepsilon_x E_{scat,x} = j\omega(\varepsilon_x - \varepsilon_o)E_{inc,x} \tag{5.5d}$$

$$\frac{\partial H_{scat,x}}{\partial z} - \frac{\partial H_{scat,z}}{\partial x} - j\omega\varepsilon_y E_{scat,y} = j\omega(\varepsilon_y - \varepsilon_o)E_{inc,y} \tag{5.5e}$$

$$\frac{\partial H_{scat,y}}{\partial x} - \frac{\partial H_{scat,x}}{\partial y} - j\omega\varepsilon_z E_{scat,z} = j\omega(\varepsilon_z - \varepsilon_o)E_{inc,z} . \tag{5.5f}$$

We will deal with 2D problems, so it is necessary to assume that there is no variation in one direction, the z direction in this case. With this in mind, equations (5.5) can be rewritten for

$\partial/\partial z = 0$ as:

$$\frac{\partial E_{scat,z}}{\partial y} + j\omega\mu_x H_{scat,x} = j\omega\left(\mu_o - \mu_x\right) H_{inc,x} \tag{5.6a}$$

$$-\frac{\partial E_{scat,z}}{\partial x} + j\omega\mu_y H_{scat,y} = j\omega\left(\mu_o - \mu_y\right) H_{inc,y} \tag{5.6b}$$

$$\frac{\partial E_{scat,y}}{\partial x} - \frac{\partial E_{scat,x}}{\partial y} + j\omega\mu_z H_{scat,z} = j\omega\left(\mu_o - \mu_z\right) H_{inc,z} \tag{5.6c}$$

$$\frac{\partial H_{scat,z}}{\partial y} - j\omega\varepsilon_x E_{scat,x} = j\omega\left(\varepsilon_x - \varepsilon_o\right) E_{inc,x} \tag{5.6d}$$

$$-\frac{\partial H_{scat,z}}{\partial x} - j\omega\varepsilon_y E_{scat,y} = j\omega\left(\varepsilon_y - \varepsilon_o\right) E_{inc,y} \tag{5.6e}$$

$$\frac{\partial H_{scat,y}}{\partial x} - \frac{\partial H_{scat,x}}{\partial y} - j\omega\varepsilon_z E_{scat,z} = j\omega\left(\varepsilon_z - \varepsilon_o\right) E_{inc,z} \,. \tag{5.6f}$$

(5.6) can be arranged to yield two sets of equations, the TM_Z equations given by:

$$H_{scat,x} + \frac{1}{j\omega\mu_x}\frac{\partial E_{scat,z}}{\partial y} = \frac{\mu_o - \mu_x}{\mu_x} H_{inc,x} \tag{5.7a}$$

$$H_{scat,y} - \frac{1}{j\omega\mu_y}\frac{\partial E_{scat,z}}{\partial x} = \frac{\mu_o - \mu_y}{\mu_y} H_{inc,y} \tag{5.7b}$$

$$E_{scat,z} + \frac{1}{j\omega\varepsilon_z}\frac{\partial H_{scat,x}}{\partial y} - \frac{1}{j\omega\varepsilon_z}\frac{\partial H_{scat,y}}{\partial x} = \frac{\varepsilon_o - \varepsilon_z}{\varepsilon_z} E_{inc,z} \,. \tag{5.7c}$$

and the TE_Z equations given by:

$$E_{scat,x} - \frac{1}{j\omega\varepsilon_x}\frac{\partial H_{scat,z}}{\partial y} = \frac{\varepsilon_o - \varepsilon_x}{\varepsilon_x} E_{inc,x} \tag{5.8a}$$

$$E_{scat,y} + \frac{1}{j\omega\varepsilon_y}\frac{\partial H_{scat,z}}{\partial x} = \frac{\varepsilon_o - \varepsilon_y}{\varepsilon_y} E_{inc,y} \tag{5.8b}$$

$$H_{scat,z} + \frac{1}{j\omega\mu_z}\frac{\partial E_{scat,y}}{\partial x} - \frac{1}{j\omega\mu_z}\frac{\partial E_{scat,x}}{\partial y} = \frac{\mu_o - \mu_z}{\mu_z} H_{inc,z} \tag{5.8c}$$

5.2 PERFECTLY MATCHED LAYER

In Chapter 4, the considered problems were all terminated by PEC boundary conditions. However, in this chapter, the problem is to calculate the scattering from a body in free space, which requires that no boundaries exist. This would require the entire space to be meshed. We cannot extend the computation space to infinity, as this would require an infinite number of cells, which is impossible.

In fact, the objective is to keep the computation space as small as possible. The smaller the computation space, the smaller the number of cells and the required CPU time and memory to perform the calculations. So, in order to make the computational space finite, the introduction of artificial boundaries that simulate the open space is necessary. The absorbing boundary conditions (ABC) simulate the open space by permitting waves propagating outward from the object to be modeled without reflection artifacts. ABC's should absorb waves with wavefronts incident at all angles and should be computationally simple.

Various absorbing boundary conditions [57, 58, 59, 60, 61] have been used in finite difference calculations, but the boundary condition called the perfectly matched layer (PML) [61, 62, 63], proposed by Berenger, provides the best performance of the bunch in terms of flexibility and efficiency. PML technique requires that the computation space is surrounded by a matched material which has both electrical and magnetic conductivities. These conductivities are specially designed to absorb electromagnetic waves without reflection for all incident angles and at any frequency. Thus, it is said to be a perfectly matched layer.

The PML technique is applied by constructing an anisotropic PML absorber just outside the original computational domain. Properties of the PML layer have been chosen to effectively absorb all outgoing waves. The extended computational domain for a 2D problem is illustrated in Figure 5.1. Maxwell's equations are solved by the FDFD or MRFD technique inside the extended computational domain which is terminated by a perfectly conducting (PEC) boundary condition. In the two dimensional case, on the left and right side of the computational domain, the absorbing layers only have nonzero conductivity in the x direction, i.e., $\sigma_y^e = 0$, $\sigma_y^m = 0$. Similarly, the conductivity in the y direction is nonzero on the top and bottom sides. In the four corners both $\sigma_x^{e,m}$ and $\sigma_y^{e,m}$ are nonzero.

The theoretical reflection from the PML layer should be zero; however some reflection may occur due to numerical discretization. To reduce this reflection, both electric and magnetic conductivities are chosen to increase from zero at the vacuum-PML interface to a value σ_{max} at the outer layer of the PML. σ_{max} can be determined from [61] as:

$$\sigma_{max} = -\frac{\varepsilon_0 c\,(n+1)\ln[R(0)]}{2\delta_{PML}} \tag{5.9}$$

with n being 1 for a linear conductivity or 2 for a parabolic conductivity profile. The parameter δ_{PML} is the PML layer thickness, c is the speed of light in vacuum, and $R(0)$ is the theoretical reflection factor at normal incidence.

The conductivity distribution inside the absorbing layers can be determined as

$$\sigma(h) = \sigma_{max}\left(\frac{h}{\delta_{PML}}\right)^n \tag{5.10}$$

where h is the distance from the vacuum-PML interface to a point inside the PML media. These relations make the PML layers absorb all waves propagating from the computational domain toward the PEC boundary.

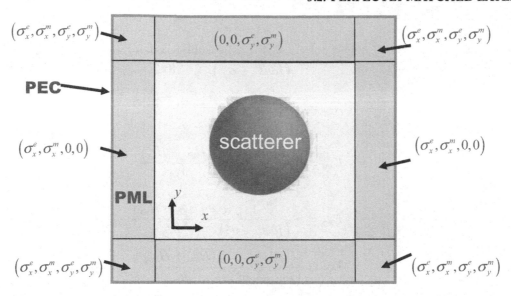

Figure 5.1: Extended computational domain with PML layers for 2-D problems.

In a split PML, Cartesian field components are split into two terms [64]. For example, a field component such as the x component of the field $\overrightarrow{F}$, denoted F_x, has been split into subfields as $F_x = F_{xy} + F_{xz}$. In three dimensions, the resulting PML modification of the time-harmonic Maxwell's equations yields twelve equations, as follows:

$$H_{xy} = -\frac{1}{\left(j\omega\mu_o + \sigma_y^m\right)}\frac{\partial\left(E_{zx} + E_{zy}\right)}{\partial y} \qquad (5.11a)$$

$$H_{xz} = \frac{1}{\left(j\omega\mu_o + \sigma_z^m\right)}\frac{\partial\left(E_{yz} + E_{yx}\right)}{\partial z} \qquad (5.11b)$$

$$H_{yz} = -\frac{1}{\left(j\omega\mu_o + \sigma_z^m\right)}\frac{\partial\left(E_{xy} + E_{xz}\right)}{\partial z} \qquad (5.11c)$$

$$H_{yx} = \frac{1}{\left(j\omega\mu_o + \sigma_x^m\right)}\frac{\partial\left(E_{zx} + E_{zy}\right)}{\partial x} \qquad (5.11d)$$

$$H_{zx} = -\frac{1}{(j\omega\mu_o + \sigma_x^m)}\frac{\partial\left(E_{yz} + E_{yx}\right)}{\partial x} \tag{5.11e}$$

$$H_{zy} = \frac{1}{\left(j\omega\mu_o + \sigma_y^m\right)}\frac{\partial\left(E_{xy} + E_{xz}\right)}{\partial y} \tag{5.11f}$$

$$E_{xy} = \frac{1}{\left(j\omega\varepsilon_o + \sigma_y^e\right)}\frac{\partial\left(H_{zx} + H_{zy}\right)}{\partial y} \tag{5.11g}$$

$$E_{xz} = -\frac{1}{(j\omega\varepsilon_o + \sigma_z^e)}\frac{\partial\left(H_{yz} + H_{yx}\right)}{\partial z} \tag{5.11h}$$

$$E_{yz} = \frac{1}{(j\omega\varepsilon_o + \sigma_z^e)}\frac{\partial\left(H_{xy} + H_{xz}\right)}{\partial z} \tag{5.11i}$$

$$E_{yx} = -\frac{1}{(j\omega\varepsilon_o + \sigma_x^e)}\frac{\partial\left(H_{zx} + H_{zy}\right)}{\partial x} \tag{5.11j}$$

$$E_{zx} = \frac{1}{(j\omega\varepsilon_o + \sigma_x^e)}\frac{\partial\left(H_{yz} + H_{yx}\right)}{\partial x} \tag{5.11k}$$

$$E_{zy} = -\frac{1}{\left(j\omega\varepsilon_o + \sigma_y^e\right)}\frac{\partial\left(H_{xy} + H_{xz}\right)}{\partial y} \tag{5.11l}$$

where μ_o and ε_o are assumed to be the free space permeability and permittivity, and σ^m and σ^e are the magnetic and electric conductivities, respectively.

The split field algorithm has some disadvantages, such as a relatively large memory requirement and the difficulty to match lossy media. In order to overcome the mentioned disadvantages, an unsplitted field algorithm [65] can be used in the time-harmonic cases. This technique is based on a Maxwellian formulation which makes the technique easier to use and more computationally efficient. According to this technique, the field components can be unified such that

$$
\begin{aligned}
E_x &= E_{xy} + E_{xz} & H_x &= H_{xy} + H_{xz} \\
E_y &= E_{yz} + E_{yx} & H_y &= H_{yz} + H_{yx} \\
E_z &= E_{zx} + E_{zy} & H_z &= H_{zx} + H_{zy} \, .
\end{aligned}
\tag{5.12}
$$

Therefore, in the time-harmonic case, certain pairs of equations, e.g., (5.11a) and (5.11b), can be added together to obtain (5.13a). This gives us an equivalent system of scalar equations called the unsplit field PML equations:

$$H_x = \frac{1}{\left(j\omega\mu_o + \sigma_z^m\right)}\frac{\partial E_y}{\partial z} - \frac{1}{\left(j\omega\mu_o + \sigma_y^m\right)}\frac{\partial E_z}{\partial y} \qquad (5.13a)$$

$$H_y = \frac{1}{\left(j\omega\mu_o + \sigma_x^m\right)}\frac{\partial E_z}{\partial x} - \frac{1}{\left(j\omega\mu_o + \sigma_z^m\right)}\frac{\partial E_x}{\partial z} \qquad (5.13b)$$

$$H_z = \frac{1}{\left(j\omega\mu_o + \sigma_y^m\right)}\frac{\partial E_x}{\partial y} - \frac{1}{\left(j\omega\mu_o + \sigma_x^m\right)}\frac{\partial E_y}{\partial x} \qquad (5.13c)$$

$$E_x = \frac{1}{\left(j\omega\varepsilon_o + \sigma_y^e\right)}\frac{\partial H_z}{\partial y} - \frac{1}{\left(j\omega\varepsilon_o + \sigma_z^e\right)}\frac{\partial H_y}{\partial z} \qquad (5.13d)$$

$$E_y = \frac{1}{\left(j\omega\varepsilon_o + \sigma_z^e\right)}\frac{\partial H_x}{\partial z} - \frac{1}{\left(j\omega\varepsilon_o + \sigma_x^e\right)}\frac{\partial H_z}{\partial x} \qquad (5.13e)$$

$$E_z = \frac{1}{\left(j\omega\varepsilon_o + \sigma_x^e\right)}\frac{\partial H_y}{\partial x} - \frac{1}{\left(j\omega\varepsilon_o + \sigma_y^e\right)}\frac{\partial H_x}{\partial y}. \qquad (5.13f)$$

2D TM$_Z$ FORMULATION

Assuming that there is no variation in the z-direction, the 2D TM$_Z$ PML equations can be obtained from (5.13):

$$H_x + \frac{1}{\left(j\omega\mu_o + \sigma_y^m\right)}\frac{\partial E_z}{\partial y} = 0 \qquad (5.14a)$$

$$H_y - \frac{1}{\left(j\omega\mu_o + \sigma_x^m\right)}\frac{\partial E_z}{\partial x} = 0 \qquad (5.14b)$$

$$E_z + \frac{1}{\left(j\omega\varepsilon_o + \sigma_y^e\right)}\frac{\partial H_x}{\partial y} - \frac{1}{\left(j\omega\varepsilon_o + \sigma_x^e\right)}\frac{\partial H_y}{\partial x} = 0. \qquad (5.14c)$$

If we compare the TM$_Z$ equations (5.14) to (5.7), it can be seen that they are similar. Instead of two separate sets of equations for the PML region and the non-PML region, we can work with one set of equations which is valid in the entirety of the computation space:

$$H_{scat,x} + \frac{1}{j\omega\mu_{xy}} \frac{\partial E_{scat,z}}{\partial y} = \frac{\mu_o - \mu_{xi}}{\mu_{xi}} H_{inc,x} \qquad (5.15a)$$

$$H_{scat,y} - \frac{1}{j\omega\mu_{yx}} \frac{\partial E_{scat,z}}{\partial x} = \frac{\mu_o - \mu_{yi}}{\mu_{yi}} H_{inc,y} \qquad (5.15b)$$

$$E_{scat,z} + \frac{1}{j\omega\varepsilon_{zy}} \frac{\partial H_{scat,x}}{\partial y} - \frac{1}{j\omega\varepsilon_{zx}} \frac{\partial H_{scat,y}}{\partial x} = \frac{\varepsilon_o - \varepsilon_{zi}}{\varepsilon_{zi}} E_{inc,z} . \qquad (5.15c)$$

The material parameters are distributed differently in the PML and non-PML regions. In the non-PML region (original computational domain) the material parameter distribution is described by:

$$
\begin{aligned}
\mu_{xy} &= \mu_x & \varepsilon_{zx} &= \varepsilon_z \\
\mu_{yx} &= \mu_y & \varepsilon_{zy} &= \varepsilon_z \\
\mu_{xi} &= \mu_x & \varepsilon_{zi} &= \varepsilon_z \\
\mu_{yi} &= \mu_y &
\end{aligned}
\qquad (5.16)
$$

while the material parameters inside the PML region are:

$$
\begin{aligned}
\mu_{xy} &= \mu_o + \frac{\sigma_y^m}{j\omega} & \varepsilon_{zx} &= \varepsilon_o + \frac{\sigma_x^e}{j\omega} \\
\mu_{yx} &= \mu_o + \frac{\sigma_x^m}{j\omega} & \varepsilon_{zy} &= \varepsilon_o + \frac{\sigma_y^e}{j\omega} \\
\mu_{xi} &= \mu_o & \varepsilon_{zi} &= \varepsilon_o \\
\mu_{yi} &= \mu_o . &
\end{aligned}
\qquad (5.17)
$$

Note that if the medium is lossy then the real part of ε represents the dielectric constant, and the imaginary part of ε represents the electric loss of the medium. Similarly, the real part of μ represents the magnetic permeability, and the imaginary part of it represents the magnetic loss of the medium.

2D TE$_Z$ FORMULATION

2D TE$_Z$ PML equations can be obtained from (5.13) assuming that there is no variation in the z-direction:

$$E_x - \frac{1}{\left(j\omega\varepsilon_o + \sigma_y^e\right)}\frac{\partial H_z}{\partial y} = 0 \tag{5.18a}$$

$$E_y + \frac{1}{\left(j\omega\varepsilon_o + \sigma_x^e\right)}\frac{\partial H_z}{\partial x} = 0 \tag{5.18b}$$

$$H_z + \frac{1}{\left(j\omega\mu_o + \sigma_x^m\right)}\frac{\partial E_y}{\partial x} - \frac{1}{\left(j\omega\mu_o + \sigma_y^m\right)}\frac{\partial E_x}{\partial y} = 0 \tag{5.18c}$$

Similar to the TM$_Z$ case, the TE$_Z$ equations (5.18) and (5.8) can be reduced to one set of equations which is valid in the entirety of the computation space:

$$E_{scat,x} - \frac{1}{j\omega\varepsilon_{xy}}\frac{\partial H_{scat,z}}{\partial y} = \frac{\varepsilon_o - \varepsilon_{xi}}{\varepsilon_{xi}}E_{inc,x} \tag{5.19a}$$

$$E_{scat,y} + \frac{1}{j\omega\varepsilon_{yx}}\frac{\partial H_{scat,z}}{\partial x} = \frac{\varepsilon_o - \varepsilon_{yi}}{\varepsilon_{yi}}E_{inc,y} \tag{5.19b}$$

$$H_{scat,z} + \frac{1}{j\omega\mu_{zx}}\frac{\partial E_{scat,y}}{\partial x} - \frac{1}{j\omega\mu_{zy}}\frac{\partial E_{scat,x}}{\partial y} = \frac{\mu_o - \mu_{zi}}{\mu_{zi}}H_{inc,z} \tag{5.19c}$$

Again, the material parameters are distributed differently in the PML and non-PML regions. In the non-PML region the material parameter distribution is given by:

$$\begin{aligned}
\varepsilon_{xy} &= \varepsilon_x & \mu_{zx} &= \mu_z \\
\varepsilon_{yx} &= \varepsilon_y & \mu_{zy} &= \mu_z \\
\varepsilon_{xi} &= \varepsilon_x & \mu_{zi} &= \mu_z \\
\varepsilon_{yi} &= \varepsilon_y
\end{aligned} \tag{5.20}$$

while the material parameters inside the PML region are:

$$\begin{aligned}
\varepsilon_{xy} &= \varepsilon_o + \frac{\sigma_y^e}{j\omega} & \mu_{zx} &= \mu_o + \frac{\sigma_x^m}{j\omega} \\
\varepsilon_{yx} &= \varepsilon_o + \frac{\sigma_x^e}{j\omega} & \mu_{zy} &= \mu_o + \frac{\sigma_y^m}{j\omega} \\
\varepsilon_{xi} &= \varepsilon_o & \mu_{zi} &= \mu_o \\
\varepsilon_{yi} &= \varepsilon_o \,.
\end{aligned} \tag{5.21}$$

5.3 SCATTERING FROM TWO-DIMENSIONAL OBJECTS

UPDATE EQUATIONS

The multiresolution frequency domain technique can be used to solve the 2D TM_Z scattering problem governed by the equations (5.15). For this purpose, on the extended PML computational domain, the field components and material properties are placed on the 2D lattice as shown in Figure 5.2.

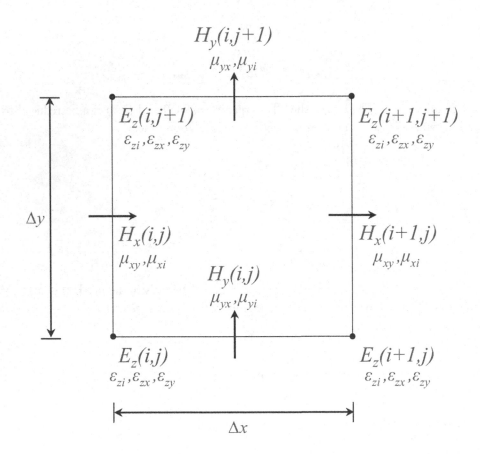

Figure 5.2: Positions of the TM_Z field components and material parameters on the 2D Yee cell.

Utilizing the MRFD scheme for TM_Z case, equations (5.15) can be discretized as:

$$H_{scat,x}(i, j) + \frac{a(1)}{j\omega\mu_{xy}(i, j)\Delta y}\left[E_{scat,z}(i, j+1) - E_{scat,z}(i, j)\right]$$

$$+ \frac{a(2)}{j\omega\mu_{xy}(i, j)\Delta y}\left[E_{scat,z}(i, j+2) - E_{scat,z}(i, j-1)\right] \qquad (5.22a)$$

$$+ \frac{a(3)}{j\omega\mu_{xy}(i, j)\Delta y}\left[E_{scat,z}(i, j+3) - E_{scat,z}(i, j-2)\right]$$

$$= \frac{\mu_o - \mu_{xi}(i, j)}{\mu_{xi}(i, j)}H_{inc,x}(i, j)$$

$$H_{scat,y}(i, j) - \frac{a(1)}{j\omega\mu_{yx}(i, j)\Delta x}\left[E_{scat,z}(i+1, j) - E_{scat,z}(i, j)\right]$$

$$- \frac{a(2)}{j\omega\mu_{yx}(i, j)\Delta x}\left[E_{scat,z}(i+2, j) - E_{scat,z}(i-1, j)\right] \qquad (5.22b)$$

$$- \frac{a(3)}{j\omega\mu_{yx}(i, j)\Delta x}\left[E_{scat,z}(i+3, j) - E_{scat,z}(i-2, j)\right]$$

$$= \frac{\mu_o - \mu_{yi}(i, j)}{\mu_{yi}(i, j)}H_{inc,y}(i, j)$$

$$- E_{scat,z}(i, j) + \frac{a(1)}{j\omega\varepsilon_{zx}(i, j)\Delta x}\left[H_{scat,y}(i, j) - H_{scat,y}(i-1, j)\right]$$

$$+ \frac{a(2)}{j\omega\varepsilon_{zx}(i, j)\Delta x}\left[H_{scat,y}(i+1, j) - H_{scat,y}(i-2, j)\right] \qquad (5.22c)$$

$$+ \frac{a(3)}{j\omega\varepsilon_{zx}(i, j)\Delta x}\left[H_{scat,y}(i+2, j) - H_{scat,y}(i-3, j)\right]$$

$$- \frac{a(1)}{j\omega\varepsilon_{zy}(i, j)\Delta y}\left[H_{scat,x}(i, j) - H_{scat,x}(i, j-1)\right]$$

$$- \frac{a(2)}{j\omega\varepsilon_{zy}(i, j)\Delta y}\left[H_{scat,x}(i, j+1) - H_{scat,x}(i, j-2)\right]$$

$$- \frac{a(3)}{j\omega\varepsilon_{zy}(i, j)\Delta y}\left[H_{scat,x}(i, j+2) - H_{scat,x}(i, j-3)\right]$$

$$= \frac{\varepsilon_{zi}(i, j) - \varepsilon_o}{\varepsilon_{zi}(i, j)}E_{inc,z}(i, j)$$

where the $a(l)$ coefficients are again given in Table 3.1.

Utilizing the 2D lattice as shown in Figure 5.3, MRFD update equations for the 2D TE_Z scattering problem governed by the equations (5.19) can be obtained:

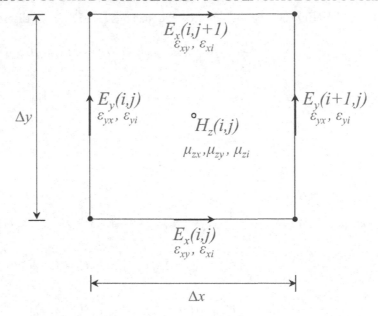

Figure 5.3: Positions of TE$_Z$ field components and material parameters on the 2D Yee cell.

$$E_{scat,x}(i, j) - \frac{a(1)}{j\omega\varepsilon_{xy}(i, j)\Delta y}\left[H_{scat,z}(i, j) - H_{scat,z}(i, j - 1)\right]$$

$$- \frac{a(2)}{j\omega\varepsilon_{xy}(i, j)\Delta y}\left[H_{scat,z}(i, j + 1) - H_{scat,z}(i, j - 2)\right] \tag{5.23a}$$

$$- \frac{a(3)}{j\omega\varepsilon_{xy}(i, j)\Delta y}\left[H_{scat,z}(i, j + 2) - H_{scat,z}(i, j - 3)\right]$$

$$= \frac{\varepsilon_o - \varepsilon_{xi}(i, j)}{\varepsilon_{xi}(i, j)}E_{inc,x}(i, j)$$

$$E_{scat,y}(i, j) + \frac{a(1)}{j\omega\varepsilon_{yx}(i, j)\Delta x}\left[H_{scat,z}(i, j) - H_{scat,z}(i - 1, j)\right]$$

$$+ \frac{a(2)}{j\omega\varepsilon_{yx}(i, j)\Delta x}\left[H_{scat,z}(i + 1, j) - H_{scat,z}(i - 2, j)\right] \tag{5.23b}$$

$$+ \frac{a(3)}{j\omega\varepsilon_{yx}(i, j)\Delta x}\left[H_{scat,z}(i + 2, j) - H_{scat,z}(i - 3, j)\right]$$

$$= \frac{\varepsilon_o - \varepsilon_{yi}(i, j)}{\varepsilon_{yi}(i, j)}E_{inc,y}(i, j)$$

$$
H_{scat,z}(i, j) + \frac{a(1)}{j\omega\mu_{zx}(i, j)\Delta x} \Big[E_{scat,y}(i + 1, j) - E_{scat,y}(i, j) \Big]
$$

$$
+ \frac{a(2)}{j\omega\mu_{zx}(i, j)\Delta x} \Big[E_{scat,y}(i + 2, j) - E_{scat,y}(i - 1, j) \Big] \tag{5.23c}
$$

$$
+ \frac{a(3)}{j\omega\mu_{zx}(i, j)\Delta x} \Big[E_{scat,y}(i + 3, j) - E_{scat,y}(i - 2, j) \Big]
$$

$$
- \frac{a(1)}{j\omega\mu_{zy}(i, j)\Delta y} \Big[E_{scat,x}(i, j + 1) - E_{scat,x}(i, j) \Big]
$$

$$
- \frac{a(2)}{j\omega\mu_{zy}(i, j)\Delta y} \Big[E_{scat,x}(i, j + 2) - E_{scat,x}(i, j - 1) \Big]
$$

$$
- \frac{a(3)}{j\omega\mu_{zy}(i, j)\Delta y} \Big[E_{scat,x}(i, j + 3) - E_{scat,x}(i, j - 2) \Big]
$$

$$
= \frac{\mu_o - \mu_{zi}(i, j)}{\mu_{zi}(i, j)} H_{inc,z}(i, j)
$$

The tangential $\vec{E}$ and normal $\vec{H}$ field components on the boundary of the extended computational boundary are set to zero (i.e., a PEC wall is assumed at the outer side of the PML layers as shown in Figure 5.1). Image principle is again used to extend the electromagnetic fields beyond the computational domain by forcing odd symmetry for tangential electric fields and normal magnetic fields. The three field components at the interior nodes can be computed by solving the linear system of equations (5.22) for the TM_Z case and (5.23) for the TE_Z case. These equations can be arranged in matrix form as $[A][EH] = [B]$ where $[A]$ is a $(N \times N)$ coefficient matrix, $[EH]$ is the unknown vector containing scattered $\vec{E}$ and $\vec{H}$ field components, and $[B]$ is the excitation vector representing the right-hand sides of (5.22) or (5.23). The excitation vector depends on all incident field components. The matrix system defined by (5.22) or (5.23) is again very sparse, since each row of the matrix has at most thirteen nonzero elements.

The parameters that are used to set up the computational domain are listed in Table 5.1.

INCIDENT FIELD EXPRESSIONS

The TM_Z incident plane wave presented in Fig. 5.4a can be expressed as:

$$
\vec{E}_{inc}(\vec{r}) = E_0 \hat{z} e^{-j\vec{k}.\vec{r}} \tag{5.24}
$$

where E_0 specifies the amplitude of the plane wave. The propagation vector $\vec{k}$ and the position vector $\vec{r}$ can be written as:

$$\vec{k} = -k(\cos \phi_{inc}\hat{x} + \sin \phi_{inc}\hat{y}) \tag{5.25a}$$

$$\vec{r} = x\hat{x} + y\hat{y} \tag{5.25b}$$

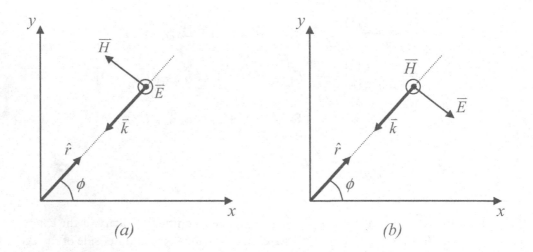

Figure 5.4: Incident plane waves in a 2D space: (a) TM_Z; (b) TE_Z.

Table 5.1: Parameters used in the extended computational domain	
Number of PML Nodes	N_{PML}
Number of Air Nodes	N_{AIR}
Number of Dielectric Nodes	N_{OBJ_X} and N_{OBJ_Y}
Number of nodes in x direction	$N_x = 2 \times (N_{PML} + N_{AIR}) + N_{OBJ\_X} + 1$
Number of nodes in y direction	$N_y = 2 \times (N_{PML} + N_{AIR}) + N_{OBJ\_Y} + 1$
Number of unknowns of E_z	$N_x \times N_y$
Number of unknowns of H_x	$N_x \times (N_y - 1)$
Number of unknowns of H_y	$N_y \times (N_x - 1)$

Since the incident field is assumed to be propagating in free space, $k = k_0 = \omega\sqrt{\mu_0\varepsilon_0}$. After performing the dot product of (5.25a) and (5.25b), (5.24) becomes:

$$\vec{E}_{inc}(\vec{r}) = E_0\hat{z}e^{jk(x\cos\phi_{inc}+y\sin\phi_{inc})} \tag{5.26}$$

The $\vec{H}$ field components can be extracted as:

$$\vec{H}_{inc}(\vec{r}) = \frac{1}{\eta}\left[\hat{k} \times \vec{E}(\vec{r})\right] \tag{5.27}$$

$$= \frac{1}{\eta}E_0\cos\phi_{inc}\hat{y}e^{jk(x\cos\phi_{inc}+y\sin\phi_{inc})} - \frac{1}{\eta}E_0\sin\phi_{inc}\hat{x}e^{jk(x\cos\phi_{inc}+y\sin\phi_{inc})}$$

The incident field components for 2D TM_Z problems are:

$$E_{inc,z} = E_0e^{jk(x\cos\phi_{inc}+y\sin\phi_{inc})} \tag{5.28a}$$

$$H_{inc,x} = -\frac{1}{\eta}E_0\sin\phi_{inc}e^{jk(x\cos\phi_{inc}+y\sin\phi_{inc})} \tag{5.28b}$$

$$H_{inc,y} = \frac{1}{\eta}E_0\cos\phi_{inc}e^{jk(x\cos\phi_{inc}+y\sin\phi_{inc})}. \tag{5.28c}$$

In a similar fashion, the incident field components for the TE_Z case shown in Fig. 5.4b can be extracted. The $\vec{H}$ field can be expressed as:

$$\vec{H}_{inc}(\vec{r}) = H_0\hat{z}e^{-j\vec{k}.\vec{r}} = H_0\hat{z}e^{jk(x\cos\phi_{inc}+y\sin\phi_{inc})} \tag{5.29}$$

The electric field components can be extracted as:

$$\vec{E}_{inc}(\vec{r}) = -\eta\left[\hat{k} \times \vec{H}_{inc}(\vec{r})\right] \tag{5.30}$$

$$= -\eta H_0\cos\phi_{inc}\hat{y}e^{jk(x\cos\phi_{inc}+y\sin\phi_{inc})} + \eta H_0\sin\phi_{inc}\hat{x}e^{jk(x\cos\phi_{inc}+y\sin\phi_{inc})}$$

The incident field components for 2D TE_Z problems are:

$$H_{inc,z} = H_0e^{jk(x\cos\phi_{inc}+y\sin\phi_{inc})} \tag{5.31a}$$

$$E_{inc,x} = \eta H_0\sin\phi_{inc}e^{jk(x\cos\phi_{inc}+y\sin\phi_{inc})} \tag{5.31b}$$

$$E_{inc,y} = -\eta H_0\cos\phi_{inc}e^{jk(x\cos\phi_{inc}+y\sin\phi_{inc})}. \tag{5.31c}$$

NUMERICAL RESULTS

In this section, the finite difference and multiresolution approaches are applied to the two-dimensional problems of scattering of a plane wave from dielectric and PEC cylinders. The fields in the computational domain and in the far field region are calculated and presented. The far field calculation procedure is summarized in Appendix D. The problems are solved with both methods for the purpose of comparison.

The source codes are written in Matlab and executed using a laptop PC equipped with a Pentium M processor at 1.6 GHz and 1 GB of memory. The coefficient matrices are stored in sparse matrices to reduce memory requirements. For all of the problems, number of air layers (N_{AIR}) and PML layers (N_{PML}) is set to 4 and 8, respectively, and a parabolic variation of conductivity with $R(0) = 10^{-17}$ is used in the PML. For each example, simulation parameters and consumed computer resources are summarized in Table 5.2.

Table 5.2: Simulation parameters and computer resources consumed by the two methods

		Cell Size (mm)	Comp. Space (cells)	Matrix Size (kbyte)	Matrix Fill Rate (%)	Time (sec)
Circular Dielectric Cylinder	FDFD	2.0	174 x 174	6757	0.0040	668.1
	MRFD	5.0	84 x 84	3733	0.0404	168.4
Square Dielectric Cylinder	FDFD	2.5	124 x 124	3466	0.0077	113.9
	MRFD	6.25	64 x 64	2154	0.0685	46.1
Dielectric and PEC Cylinders	FDFD	2.5	424 x 124	11873	0.0023	2473.1
	MRFD	6.25	184 x 64	6295	0.0248	666.4

First, the problem of scattering from a circular dielectric cylinder with $\varepsilon_r = 4$, illustrated in Figure 5.5b, is considered. The cylinder is illuminated by an incident 3 GHz TM_Z plane wave with 180° incidence angle off the x-axis. The MRFD sampling rate ($\Delta x = \Delta y = 5$ mm $= \lambda_{min}/10$) is kept low compared to the higher one in the FDFD ($\Delta x = \Delta y = 2$ mm $= \lambda_{min}/25$). The magnitude of H_y along $y = 0$ is calculated and plotted in Figure 5.6. The far field distribution for the problem is also calculated. The scattered far field represented by bistatic echo width is shown in Figure 5.7. Both graphs show that there is a good agreement between MRFD and FDFD results. For this case, the memory and processing time savings of the multiresolution technique are 44.8% and 74.8%, respectively.

The second example considered is the problem of scattering from a square dielectric cylinder with $\varepsilon_r = 4$, depicted in Figure 5.5a. The width of the square cylinder is 25 cm. The cylinder is illuminated by a 3 GHz incident TE_Z plane wave with 0° incidence angle off the x-axis. Compared to the FDFD grid with ($\Delta x = \Delta y = 2.5$ mm $= \lambda_{min}/20$), the MRFD grid is coarser ($\Delta x = \Delta y = 6.25$ mm $= \lambda_{min}/8$), where λ_{min} is the wavelength inside the scatterer. The scattered co-polarized far field, represented by bistatic echo width, is presented in Figure 5.8. Both methods yield similar results, however compared to the finite difference technique, memory and processing time savings of the multiresolution technique are 37.8% and 59.6%, respectively, for the case at hand.

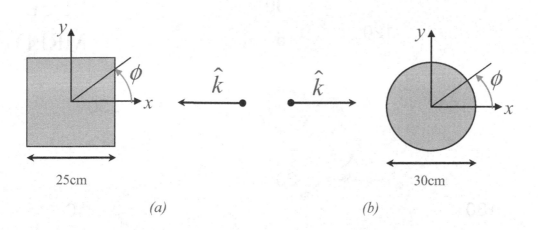

Figure 5.5: 2D dielectric cylinders illuminated by plane waves: (a) square cylinder; (b) circular cylinder.

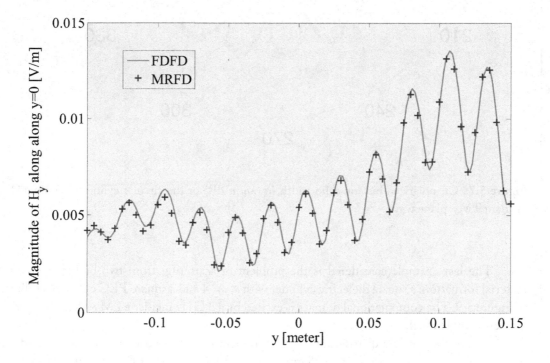

Figure 5.6: Magnitude of H_y along $y = 0$, scattering from circular cylinder.

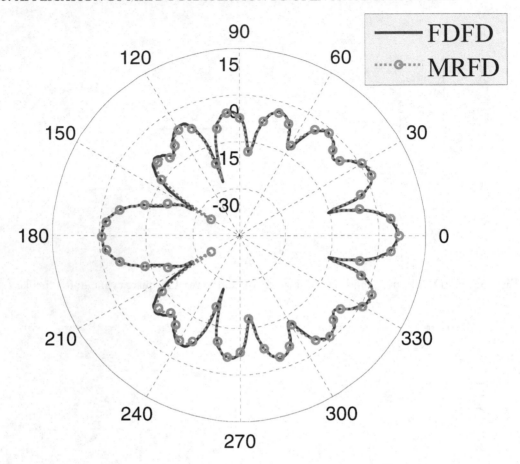

Figure 5.7: Co-polarized bistatic echo width (σ/λ_0 in dB) of the circular cylinder illuminated by an incident TM_Z plane wave.

The last example considered is the problem of scattering from two bodies with different material formation: a square dielectric cylinder with $\varepsilon_r = 4$ and a square PEC cylinder, as illustrated in Figure 5.9. The computational domain is excited by a 3 GHz incident TM_Z plane wave with 90° incidence angle off the x-axis.

Again, the far field distribution for the problem is calculated with the MRFD and FDFD techniques. The cell sizes are set to 6.25 mm and 2.5 mm for the MRFD and FDFD lattice, respectively. The scattered co-polarized far field for the problem is shown in Figure 5.10. Both methods yielding comparable accuracy, memory, and simulation time requirements of the multiresolution technique is 47% and 73% lower than the FDFD.

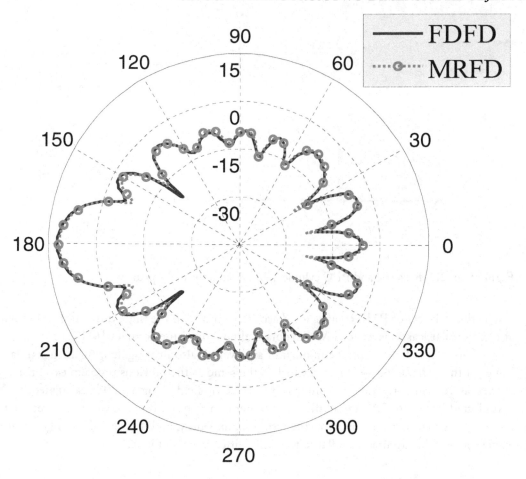

Figure 5.8: Co-polarized bistatic echo width (σ/λ_0 in dB) of the square cylinder illuminated by an incident TE_Z plane wave.

Besides the grid resolution, the numbers of air layers and PML layers also have an effect on the efficiency of the method. The effect of changing the number of PML layers on simulation accuracy is investigated. For this purpose, the square dielectric cylinder shown in Figure 5.5a is considered. The cylinder is illuminated by a 3 GHz incident TM_Z plane wave with $0°$ incidence angle off the x-axis. First, a standard set of values of the parameters are selected. The number of air layers is chosen to be 16, cell size is chosen to be 5 mm, and a parabolic variation of conductivity in the PML with $R(0) = 10^{-17}$ is chosen. The magnitude of total electric field along $x = 0$ is calculated via the MRFD scheme for different number of PML layers and plotted in Figure 5.11. It is clear from the

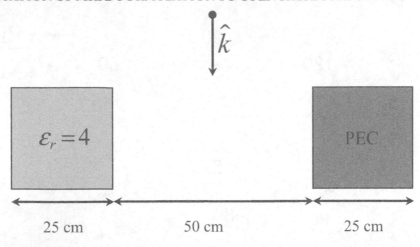

Figure 5.9: 2D dielectric and PEC cylinders illuminated by TM_Z plane wave.

results that 8 layers of PML provide good accuracy, similar to the finite difference technique which also requires at least 8 layers of PML to obtain almost the same accuracy [68].

The effect of changing the number of air layers is also investigated. Again, a standard set of values of the parameters are selected which is the same as the previous problem except that N_{PML} is fixed at 16. Similarly, the magnitude of total electric field along $x = 0$ is calculated via MRFD and plotted in Figure 5.12 for a different number of air layers. It is clear from the results that the number of air layers does not affect the results significantly, and 4 layers of air is good enough to achieve acceptable accuracy, similar to the finite difference technique.

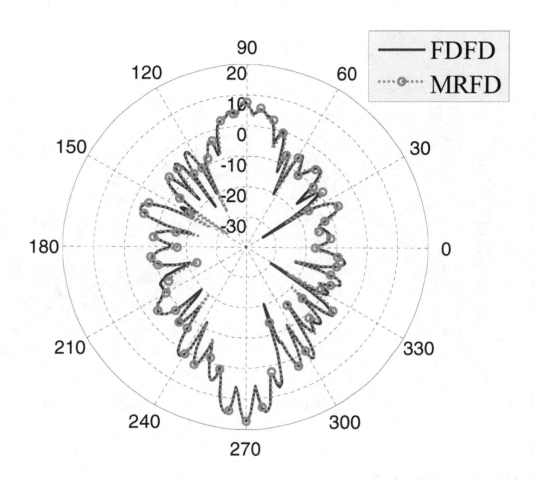

Figure 5.10: Co-polarized bistatic echo width (σ/λ_0 in dB) for the dielectric PEC square cylinders illuminated by an incident TM_Z plane wave.

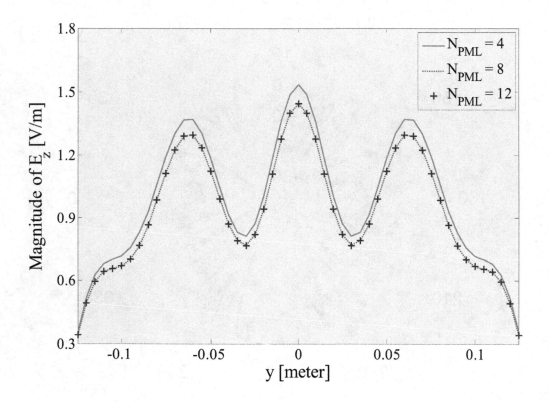

Figure 5.11: Magnitude of E_Z along $x = 0$ for various N_{PML}.

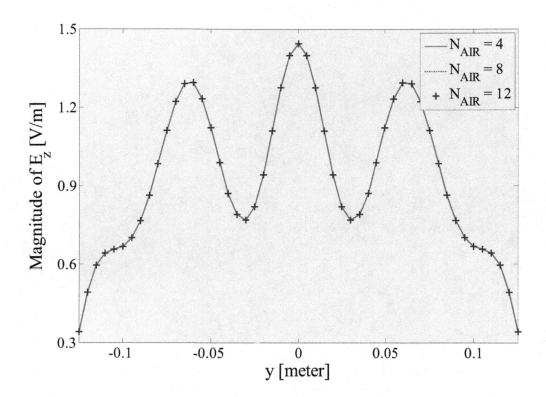

Figure 5.12: Magnitude of E_Z along $x = 0$ for various N_{AIR}.

CHAPTER 6

A Multiresolution Frequency Domain Formulation for Inhomogeneous Media

Multiresolution schemes for modeling homogeneous mediums are introduced in Chapter 3. In this chapter, an MRFD scheme for analyzing inhomogeneous problems is formulated.

6.1 DERIVATION OF THE INHOMOGENEOUS MULTIRESOLUTION FREQUENCY DOMAIN SCHEME

Multiresolution frequency domain techniques for modeling problems with homogeneous media are formulated in Chapter 3; however, these formulations are easily extended to apply to problems with inhomogeneous media. Satisfactory results are achieved with the aid of pointwise sampling of material properties and averaging of material parameters at the interfaces between neighboring mediums. These approaches introduce some errors which may increase if the contrast ratio of the material properties of the neighboring mediums is high. In this section, we formulate a CDF(2,2) wavelet based MRFD technique to model inhomogeneous mediums in a mathematically correct way, which from now on we call the inhomogeneous MRFD formulation. In addition, magnetic and electric conductivities are taken into consideration in the Maxwell's curl equations for a more general approach.

Considered Maxwell's curl equations in matrix form are:

$$
\begin{bmatrix}
0 & -\partial/\partial z & \partial/\partial y \\
\partial/\partial z & 0 & -\partial/\partial x \\
-\partial/\partial y & \partial/\partial x & 0
\end{bmatrix}
\begin{bmatrix}
H_x \\
H_y \\
H_z
\end{bmatrix}
\tag{6.1a}
$$

$$
= j\omega
\begin{bmatrix}
\varepsilon_x & 0 & 0 \\
0 & \varepsilon_y & 0 \\
0 & 0 & \varepsilon_z
\end{bmatrix}
\begin{bmatrix}
E_x \\
E_y \\
E_z
\end{bmatrix}
+
\begin{bmatrix}
\sigma_x^e & 0 & 0 \\
0 & \sigma_y^e & 0 \\
0 & 0 & \sigma_z^e
\end{bmatrix}
\begin{bmatrix}
E_x \\
E_y \\
E_z
\end{bmatrix}
$$

$$
\begin{bmatrix}
0 & \partial/\partial z & -\partial/\partial y \\
-\partial/\partial z & 0 & \partial/\partial x \\
\partial/\partial y & -\partial/\partial x & 0
\end{bmatrix}
\begin{bmatrix}
E_x \\
E_y \\
E_z
\end{bmatrix}
\tag{6.1b}
$$

$$
= j\omega
\begin{bmatrix}
\mu_x & 0 & 0 \\
0 & \mu_y & 0 \\
0 & 0 & \mu_z
\end{bmatrix}
\begin{bmatrix}
H_x \\
H_y \\
H_z
\end{bmatrix}
+
\begin{bmatrix}
\sigma_x^m & 0 & 0 \\
0 & \sigma_y^m & 0 \\
0 & 0 & \sigma_z^m
\end{bmatrix}
\begin{bmatrix}
H_x \\
H_y \\
H_z
\end{bmatrix}.
$$

These curl equations lead to six scalar equations in a Cartesian coordinate system:

$$
j\omega\varepsilon_x(x,y,z)E_x(x,y,z) + \sigma_x^e(x,y,z)E_x(x,y,z) = \frac{\partial H_z(x,y,z)}{\partial y} - \frac{\partial H_y(x,y,z)}{\partial z}
\tag{6.2a}
$$

$$
j\omega\varepsilon_y(x,y,z)E_y(x,y,z) + \sigma_y^e(x,y,z)E_y(x,y,z) = \frac{\partial H_x(x,y,z)}{\partial z} - \frac{\partial H_z(x,y,z)}{\partial x}
\tag{6.2b}
$$

$$
j\omega\varepsilon_z(x,y,z)E_z(x,y,z) + \sigma_z^e(x,y,z)E_z(x,y,z) = \frac{\partial H_y(x,y,z)}{\partial x} - \frac{\partial H_x(x,y,z)}{\partial y}
\tag{6.2c}
$$

$$
j\omega\mu_x(x,y,z)H_x(x,y,z) + \sigma_x^m(x,y,z)H_x(x,y,z) = -\frac{\partial E_z(x,y,z)}{\partial y} + \frac{\partial E_y(x,y,z)}{\partial z}
\tag{6.2d}
$$

$$
j\omega\mu_y(x,y,z)H_y(x,y,z) + \sigma_y^m(x,y,z)H_y(x,y,z) = -\frac{\partial E_x(x,y,z)}{\partial z} + \frac{\partial E_z(x,y,z)}{\partial x}
\tag{6.2e}
$$

$$
j\omega\mu_z(x,y,z)H_z(x,y,z) + \sigma_z^m(x,y,z)H_z(x,y,z) = -\frac{\partial E_y(x,y,z)}{\partial x} + \frac{\partial E_x(x,y,z)}{\partial y}.
\tag{6.2f}
$$

For the discretization of these equations by the method of moments, the field components should be first expanded in terms of the basis functions, which for this case is the dual scaling function of the CDF(2,2) wavelet base:

$$
E_x(x,y,z) = \sum_{i',j',k'} E_x(i',j',k')\tilde{\phi}_{i'+1/2}(x)\tilde{\phi}_{j'}(y)\tilde{\phi}_{k'}(z)
\tag{6.3a}
$$

$$
E_y(x,y,z) = \sum_{i',j',k'} E_y(i',j',k')\tilde{\phi}_{i'}(x)\tilde{\phi}_{j'+1/2}(y)\tilde{\phi}_{k'}(z)
\tag{6.3b}
$$

$$
E_z(x,y,z) = \sum_{i',j',k'} E_z(i',j',k')\tilde{\phi}_{i'}(x)\tilde{\phi}_{j'}(y)\tilde{\phi}_{k'+1/2}(z)
\tag{6.3c}
$$

$$H_x(x, y, z) = \sum_{i', j', k'} H_x(i', j', k')\tilde{\phi}_{i'}(x)\tilde{\phi}_{j'+1/2}(y)\tilde{\phi}_{k'+1/2}(z) \qquad (6.3d)$$

$$H_y(x, y, z) = \sum_{i', j', k'} H_y(i', j', k')\tilde{\phi}_{i'+1/2}(x)\tilde{\phi}_{j'}(y)\tilde{\phi}_{k'+1/2}(z) \qquad (6.3e)$$

$$H_z(x, y, z) = \sum_{i', j', k'} H_z(i', j', k')\tilde{\phi}_{i'+1/2}(x)\tilde{\phi}_{j'+1/2}(y)\tilde{\phi}_{k'}(z) . \qquad (6.3f)$$

Like the homogeneous case, the indexes i', j', k' indicate the discrete space lattice related to the space grid through $x = i'\Delta x$, $y = j'\Delta y$, and $z = k'\Delta z$, and the locations of the scaling functions are selected such that they are centered at the location of the corresponding field component on the Yee cell.

The procedure for deriving the update equations of all six scalar equations is similar. Therefore, for the sake of illustration, we derive the update equations only for the x components of the electric and magnetic fields.

The update equation for (6.2a) is considered first. In order to obtain relatively simple update equations, we define the functions $\varepsilon_{rx}(x, y, z)$ and $\sigma_x^e(x, y, z)$ following the procedure of [67], such that:

$$\varepsilon_x(x, y, z) = \varepsilon_0 [1 + \varepsilon_{rx}(x, y, z)] \qquad (6.4a)$$

$$\varepsilon_{rx}(x, y, z) = \varepsilon_{rx}(x)\varepsilon_{rx}(y)\varepsilon_{rx}(z) \qquad (6.4b)$$

$$\sigma_x^e(x, y, z) = \sigma_x^e(x)\sigma_x^e(y)\sigma_x^e(z). \qquad (6.4c)$$

The outline of the derivation is similar to the homogeneous case. The field expansions (6.3) are substituted into (6.2a) and tested with $\phi_{i+1/2}(x)\phi_j(y)\phi_k(z)$, following the Galerkin's method. The first term of the left-hand side of (6.2a) is tested as follows:

$$j\omega \int_{-\infty}^{\infty} \int_{-\infty}^{\infty} \int_{-\infty}^{\infty} \varepsilon_x(x, y, z)E_x(x, y, z)\phi_{i+1/2}(x)\phi_j(y)\phi_k(z)dxdydz$$

$$= j\omega\varepsilon_0 \int_{-\infty}^{\infty} \int_{-\infty}^{\infty} \int_{-\infty}^{\infty} [1 + \varepsilon_{rx}(x, y, z)] E_x(x, y, z)\phi_{i+1/2}(x)\phi_j(y)\phi_k(z)dxdydz$$

$$= j\omega\varepsilon_0 \int_{-\infty}^{\infty}\int_{-\infty}^{\infty}\int_{-\infty}^{\infty} \left[\sum_{i',j',k'} E_x(i',j',k')\tilde{\phi}_{i'+1/2}(x)\tilde{\phi}_{j'}(y)\tilde{\phi}_{k'}(z) \right] \phi_{i+1/2}(x)\phi_j(y)\phi_k(z)dxdydz$$

$$+ j\omega\varepsilon_0 \int_{-\infty}^{\infty}\int_{-\infty}^{\infty}\int_{-\infty}^{\infty} \left(\varepsilon_{rx}(x,y,z) \left[\sum_{i',j',k'} E_x(i',j',k')\tilde{\phi}_{i'+1/2}(x)\tilde{\phi}_{j'}(y)\tilde{\phi}_{k'}(z) \right] \right.$$
$$\left. \cdot\phi_{i+1/2}(x)\phi_j(y)\phi_k(z)dxdydz \right)$$

$$= j\omega\varepsilon_0 \sum_{i',j',k'} E_x(i',j',k') \int_{-\infty}^{\infty} \tilde{\phi}_{i'+1/2}(x)\phi_{i+1/2}(x)dx \int_{-\infty}^{\infty} \tilde{\phi}_{j'}(y)\phi_j(y)dy \int_{-\infty}^{\infty} \tilde{\phi}_{k'}(z)\phi_k(z)dz$$

$$+ j\omega\varepsilon_0 \sum_{i',j',k'} \tag{6.5}$$

$$\left(\begin{array}{l} E_x(i',j',k')\cdot \\ \int_{-\infty}^{\infty} \varepsilon_{rx}(x)\tilde{\phi}_{i'+1/2}(x)\phi_{i+1/2}(x)dx \int_{-\infty}^{\infty} \varepsilon_{rx}(y)\tilde{\phi}_{j'}(y)\phi_j(y)dy \int_{-\infty}^{\infty} \varepsilon_{rx}(z)\tilde{\phi}_{k'}(z)\phi_k(z)dz \end{array} \right)$$

$$= j\omega\varepsilon_0 E_x(i,j,k)\Delta x \Delta y \Delta z + j\omega\varepsilon_0 \Delta x \Delta y \Delta z \sum_{i',j',k'} \left(E_x(i',j',k')\varepsilon_{rx}^{i'+1/2,i+1/2}\varepsilon_{rx}^{j',j}\varepsilon_{rx}^{k',k} \right)$$

where

$$\varepsilon_{rx}^{i'+1/2,i+1/2} = \frac{1}{\Delta x}\int_{-\infty}^{\infty} \varepsilon_{rx}(x)\tilde{\phi}_{i'+1/2}(x)\phi_{i+1/2}(x)dx \tag{6.6a}$$

$$\varepsilon_{rx}^{j',j} = \frac{1}{\Delta y}\int_{-\infty}^{\infty} \varepsilon_{rx}(y)\tilde{\phi}_{j'}(y)\phi_j(y)dy \tag{6.6b}$$

$$\varepsilon_{rx}^{k',k} = \frac{1}{\Delta z}\int_{-\infty}^{\infty} \varepsilon_{rx}(z)\tilde{\phi}_{k'}(z)\phi_k(z)dz. \tag{6.6c}$$

Next, the second term of the left-hand side of (6.2a) is tested:

$$\int\limits_{-\infty}^{\infty}\int\limits_{-\infty}^{\infty}\int\limits_{-\infty}^{\infty}\sigma_x^e(x,y,z)E_x(x,y,z)\phi_{i+1/2}(x)\phi_j(y)\phi_k(z)dxdydz \tag{6.7}$$

$$=\int\limits_{-\infty}^{\infty}\int\limits_{-\infty}^{\infty}\int\limits_{-\infty}^{\infty}\sigma_x^e(x,y,z)\left[\sum_{i',j',k'}E_x(i',j',k')\tilde{\phi}_{i'+1/2}(x)\tilde{\phi}_{j'}(y)\tilde{\phi}_{k'}(z)\right]\phi_{i+1/2}(x)\phi_j(y)\phi_k(z)dxdydz$$

$$=\sum_{i',j',k'}E_x(i',j',k')\int\limits_{-\infty}^{\infty}\int\limits_{-\infty}^{\infty}\int\limits_{-\infty}^{\infty}\sigma_x^e(x,y,z)\tilde{\phi}_{i'+1/2}(x)\tilde{\phi}_{j'}(y)\tilde{\phi}_{k'}(z)\phi_{i+1/2}(x)\phi_j(y)\phi_k(z)dxdydz$$

$$=\sum_{i',j',k'}E_x(i',j',k')\int\limits_{-\infty}^{\infty}\sigma_x^e(x)\tilde{\phi}_{i'+1/2}(x)\phi_{i+1/2}(x)dx\int\limits_{-\infty}^{\infty}\sigma_x^e(y)\tilde{\phi}_{j'}(y)\phi_j(y)dy$$

$$\int\limits_{-\infty}^{\infty}\sigma_x^e(z)\tilde{\phi}_{k'}(z)\phi_k(z)dz$$

$$=\Delta x\Delta y\Delta z\sum_{i',j',k'}E_x(i',j',k')\sigma_{ex}^{i'+1/2,i+1/2}\sigma_{ex}^{j',j}\sigma_{ex}^{k',k}$$

where

$$\sigma_{ex}^{i'+1/2,i+1/2}=\frac{1}{\Delta x}\int\limits_{-\infty}^{\infty}\sigma_x^e(x)\tilde{\phi}_{i'+1/2}(x)\phi_{i+1/2}(x)dx \tag{6.8a}$$

$$\sigma_{ex}^{j',j}=\frac{1}{\Delta y}\int\limits_{-\infty}^{\infty}\sigma_x^e(y)\tilde{\phi}_{j'}(y)\phi_j(y)dy \tag{6.8b}$$

$$\sigma_{ex}^{k',k}=\frac{1}{\Delta z}\int\limits_{-\infty}^{\infty}\sigma_x^e(z)\tilde{\phi}_{k'}(z)\phi_k(z)dz. \tag{6.8c}$$

Then, the first term of the right-hand side of (6.2a) is tested:

$$\int_{-\infty}^{\infty}\int_{-\infty}^{\infty}\int_{-\infty}^{\infty}\frac{\partial H_z(x,y,z)}{\partial y}\phi_{i+1/2}(x)\phi_j(y)\phi_k(z)dxdydz \tag{6.9}$$

$$=\int_{-\infty}^{\infty}\int_{-\infty}^{\infty}\int_{-\infty}^{\infty}\frac{\partial}{\partial y}\left[\sum_{i,j,k}H_z(i',j',k')\tilde{\phi}_{i'+1/2}(x)\tilde{\phi}_{j'+1/2}(y)\tilde{\phi}_{k'}(z)\right]$$

$$\phi_{i+1/2}(x)\phi_j(y)\phi_k(z)dxdydz$$

$$=\sum_{i',j',k'}H_z(i',j',k')\int_{-\infty}^{\infty}\tilde{\phi}_{i'+1/2}(x)\phi_{i+1/2}(x)dx\int_{-\infty}^{\infty}\frac{\partial\tilde{\phi}_{j'+1/2}(y)}{\partial y}\phi_j(y)dy\int_{-\infty}^{\infty}\tilde{\phi}_{k'}(z)\phi_k(z)dz$$

$$=\Delta x\Delta z\sum_{l=1}^{3}a(l)\left[H_z(i,j+l-1,k)-H_z(i,j-l,k)\right].$$

Finally, the second term of the right-hand side of (6.2a) can similarly be sampled to yield:

$$\int_{-\infty}^{\infty}\int_{-\infty}^{\infty}\int_{-\infty}^{\infty}\frac{\partial H_y(x,y,z)}{\partial z}\phi_{i+1/2}(x)\phi_j(y)\phi_k(z)dxdydz \tag{6.10}$$

$$=\Delta x\Delta y\sum_{l=1}^{3}a(l)\left[H_y(i,j,k+l-1)-H_y(i,j,k-l)\right].$$

Using (6.5), (6.7), (6.9), and (6.10), the tested (6.2a) becomes the MRFD update equation for E_x:

$$j\omega\varepsilon_0\sum_{i',j',k'}E_x(i',j',k')\left[\delta_{i'+1/2,i+1/2}\delta_{j',j}\delta_{k',k}+\varepsilon_{rx}^{i'+1/2,i+1/2}\varepsilon_{rx}^{j',j}\varepsilon_{rx}^{k',k}\right] \tag{6.11}$$

$$+\sum_{i',j',k'}E_x(i',j',k')\sigma_{ex}^{i'+1/2,i+1/2}\sigma_{ex}^{j',j}\sigma_{ex}^{k',k}$$

$$=\sum_{l=1}^{3}a(l)\frac{H_z(i,j+l-1,k)-H_z(i,j-l,k)}{\Delta y}-\sum_{l=1}^{3}a(l)\frac{H_y(i,j,k+l-1)-H_y(i,j,k-l)}{\Delta z}.$$

Now, we consider the update equation for H_x. The derivation is conducted by substituting the field expansions (6.3) into (6.2d) and testing with $\phi_i(x)\phi_{j+1/2}(y)\phi_{k+1/2}(z)$, according to MoM.

We define the functions $\mu_{rx}(x, y, z)$ and $\sigma_x^m(x, y, z)$ to obtain simple update equations, such that

$$\mu_x(x, y, z) = \mu_0\left[1 + \mu_{rx}(x, y, z)\right] \tag{6.12a}$$

$$\mu_{rx}(x, y, z) = \mu_{rx}(x)\mu_{rx}(y)\mu_{rx}(z) \tag{6.12b}$$

$$\sigma_x^m(x, y, z) = \sigma_x^m(x)\sigma_x^m(y)\sigma_x^m(z). \tag{6.12c}$$

First, we test the first term of left-hand side of (6.2d):

$$j\omega \int_{-\infty}^{\infty} \int_{-\infty}^{\infty} \int_{-\infty}^{\infty} \mu_x(x, y, z)H_x(x, y, z)\phi_i(x)\phi_{j+1/2}(y)\phi_{k+1/2}(z)dxdydz \tag{6.13}$$

$$= j\omega\mu_0 \int_{-\infty}^{\infty} \int_{-\infty}^{\infty} \int_{-\infty}^{\infty} \left[1 + \mu_{rx}(x, y, z)\right]H_x(x, y, z)\phi_i(x)\phi_{j+1/2}(y)\phi_{k+1/2}(z)dxdydz$$

$$= j\omega\mu_0 \int_{-\infty}^{\infty} \int_{-\infty}^{\infty} \int_{-\infty}^{\infty} \left[\sum_{i',j',k'} H_x(i', j', k')\tilde{\phi}_{i'}(x)\tilde{\phi}_{j'+1/2}(y)\tilde{\phi}_{k'+1/2}(z)\right]$$

$$\phi_i(x)\phi_{j+1/2}(y)\phi_{k+1/2}(z)dxdydz$$

$$+ j\omega\mu_0 \int_{-\infty}^{\infty} \int_{-\infty}^{\infty} \int_{-\infty}^{\infty} \left(\mu_{rx}(x, y, z)\left[\sum_{i',j',k'} H_x(i', j', k')\tilde{\phi}_{i'}(x)\tilde{\phi}_{j'+1/2}(y)\tilde{\phi}_{k'+1/2}(z)\right]\right.$$
$$\left. \cdot \phi_i(x)\phi_{j+1/2}(y)\phi_{k+1/2}(z)dxdydz\right)$$

$$= j\omega\mu_0 \sum_{i',j',k'} H_x(i', j', k') \int_{-\infty}^{\infty} \tilde{\phi}_{i'}(x)\phi_i(x)dx \int_{-\infty}^{\infty} \tilde{\phi}_{j'+1/2}(y)\phi_{j+1/2}(y)dy$$

$$\int_{-\infty}^{\infty} \tilde{\phi}_{k'+1/2}(z)\phi_{k+1/2}(z)dz$$

$$+ j\omega\mu_0 \sum_{i',j',k'} \left(\begin{array}{c} H_x(i', j', k') \int_{-\infty}^{\infty} \mu_{rx}(x)\tilde{\phi}_{i'}(x)\phi_i(x)dx \\ \int_{-\infty}^{\infty} \mu_{rx}(y)\tilde{\phi}_{j'+1/2}(y)\phi_{j+1/2}(y)dy \int_{-\infty}^{\infty} \mu_{rx}(z)\tilde{\phi}_{k'+1/2}(z)\phi_{k+1/2}(z)dz \end{array}\right)$$

$$= j\omega\mu_0 H_x(i, j, k)\Delta x\Delta y\Delta z + j\omega\mu_0\Delta x\Delta y\Delta z$$

$$\sum_{i',j',k'} H_x(i', j', k')\mu_{rx}^{i',i}\mu_{rx}^{j'+1/2,j+1/2}\mu_{rx}^{k'+1/2,k+1/2}$$

where

$$\mu_{rx}^{i',i} = \frac{1}{\Delta x} \int_{-\infty}^{\infty} \mu_{rx}(x)\tilde{\phi}_{i'}(x)\phi_i(x)dx \tag{6.14a}$$

$$\mu_{rx}^{j'+1/2,j+1/2} = \frac{1}{\Delta y} \int_{-\infty}^{\infty} \mu_{rx}(y)\tilde{\phi}_{j'+1/2}(y)\phi_{j+1/2}(y)dy \tag{6.14b}$$

$$\mu_{rx}^{k'+1/2,k+1/2} = \frac{1}{\Delta z} \int_{-\infty}^{\infty} \mu_{rx}(z)\tilde{\phi}_{k'+1/2}(z)\phi_{k+1/2}(z)dz \ . \tag{6.14c}$$

Testing the second term of the left-hand side of (6.2d) yields:

$$\int_{-\infty}^{\infty}\int_{-\infty}^{\infty}\int_{-\infty}^{\infty} \sigma_x^m(x,y,z)H_x(x,y,z)\phi_i(x)\phi_{j+1/2}(y)\phi_{k+1/2}(z)dxdydz \tag{6.15}$$

$$= \int_{-\infty}^{\infty}\int_{-\infty}^{\infty}\int_{-\infty}^{\infty} \left(\begin{array}{l} \sigma_x^m(x,y,z) \\ \left[\sum_{i',j',k'} H_x(i',j',k')\tilde{\phi}_{i'}(x)\tilde{\phi}_{j'+1/2}(y)\tilde{\phi}_{k'+1/2}(z) \right] \phi_i(x)\phi_{j+1/2}(y)\phi_{k+1/2}(z)dxdydz \end{array} \right)$$

$$= \sum_{i',j',k'} \left(\begin{array}{l} H_x(i',j',k') \int_{-\infty}^{\infty} \sigma_x^m(x)\tilde{\phi}_{i'}(x)\phi_i(x)dx \\ \int_{-\infty}^{\infty} \sigma_x^m(y)\tilde{\phi}_{j'+1/2}(y)\phi_{j+1/2}(y)dy \int_{-\infty}^{\infty} \sigma_x^m(z)\tilde{\phi}_{k'+1/2}(z)\phi_{k+1/2}(z)dz \end{array} \right)$$

$$= \Delta x \Delta y \Delta z \sum_{i',j',k'} H_x(i',j',k')\sigma_{mx}^{i',i}\sigma_{mx}^{j'+1/2,j+1/2}\sigma_{mx}^{k'+1/2,k+1/2}$$

where

$$\sigma_{mx}^{i',i} = \frac{1}{\Delta x} \int_{-\infty}^{\infty} \sigma_x^m(x)\tilde{\phi}_{i'}(x)\phi_i(x)dx \tag{6.16a}$$

$$\sigma_{mx}^{j'+1/2,j+1/2} = \frac{1}{\Delta y} \int_{-\infty}^{\infty} \sigma_x^m(y)\tilde{\phi}_{j'+1/2}(y)\phi_{j+1/2}(y)dy \tag{6.16b}$$

$$\sigma_{mx}^{k'+1/2,k+1/2} = \frac{1}{\Delta z} \int_{-\infty}^{\infty} \sigma_x^m(z)\tilde{\phi}_{k'+1/2}(z)\phi_{k+1/2}(z)dz. \tag{6.16c}$$

Testing the first term of the right-hand side of (6.2d) yields:

$$\int_{-\infty}^{\infty}\int_{-\infty}^{\infty}\int_{-\infty}^{\infty}\frac{\partial E_z(x,y,z)}{\partial y}\phi_i(x)\phi_{j+1/2}(y)\phi_{k+1/2}(z)dxdydz \tag{6.17}$$

$$=\int_{-\infty}^{\infty}\int_{-\infty}^{\infty}\int_{-\infty}^{\infty}\frac{\partial}{\partial y}\left[\sum_{i',j',k'}E_z(i',j',k')\tilde{\phi}_{i'}(x)\tilde{\phi}_{j'}(y)\tilde{\phi}_{k'+1/2}(z)\right]\phi_i(x)\phi_{j+1/2}(y)\phi_{k+1/2}(z)dxdydz$$

$$=\sum_{i',j',k'}E_z(i',j',k')\int_{-\infty}^{\infty}\tilde{\phi}_{i'}(x)\phi_i(x)dx\int_{-\infty}^{\infty}\frac{\partial\tilde{\phi}_{j'}(y)}{\partial y}\phi_{j+1/2}(y)dy\int_{-\infty}^{\infty}\tilde{\phi}_{k'+1/2}(z)\phi_{k+1/2}(z)dz$$

$$=\Delta x\Delta z\sum_{l=1}^{3}a(l)\left[E_z(i,j+l,k)-E_z(i,j-l+1,k)\right].$$

Finally, testing the second term of the right-hand side of (6.2d) similarly yields:

$$\int_{-\infty}^{\infty}\int_{-\infty}^{\infty}\int_{-\infty}^{\infty}\frac{\partial E_y(x,y,z)}{\partial z}\phi_i(x)\phi_{j+1/2}(y)\phi_{k+1/2}(z)dxdydz \tag{6.18}$$

$$=\Delta x\Delta y\sum_{l=1}^{3}a(l)\left[E_y(i,j,k+l)-E_y(i,j,k-l+1)\right].$$

Use of (6.13), (6.15), (6.17), and (6.18) in the tested (6.2d) results in the MRFD update equation for H_x:

$$j\omega\mu_0\sum_{i',j',k'}H_x(i',j',k')\left[\delta_{i',i}\delta_{j'+1/2,j+1/2}\delta_{k'+1/2,k+1/2}+\mu_{rx}^{i',i}\mu_{rx}^{j'+1/2,j+1/2}\mu_{rx}^{k'+1/2,k+1/2}\right]$$

$$+\sum_{i',j',k'}H_x(i',j',k')\sigma_{mx}^{i',i}\sigma_{mx}^{j'+1/2,j+1/2}\sigma_{mx}^{k'+1/2,k+1/2} \tag{6.19}$$

$$=-\sum_{l=1}^{3}a(l)\frac{E_z(i,j+l,k)-E_z(i,j-l+1,k)}{\Delta y}+\sum_{l=1}^{3}a(l)\frac{E_y(i,j,k+l)-E_y(i,j,k-l+1)}{\Delta z}.$$

The update equations for the rest of the field components can be developed by following the same procedure. These equations are summarized as:

$$
\begin{aligned}
&\left(
\begin{aligned}
& j\omega\varepsilon_0 \sum_{i',j',k'} E_x(i',j',k') \left[\delta_{i'+1/2,i+1/2}\delta_{j',j}\delta_{k',k} + \varepsilon_{rx}^{i'+1/2,i+1/2}\varepsilon_{rx}^{j',j}\varepsilon_{rx}^{k',k} \right] \\
& \quad + \sum_{i',j',k'} E_x(i',j',k')\sigma_{ex}^{i'+1/2,i+1/2}\sigma_{ex}^{j',j}\sigma_{ex}^{k',k}
\end{aligned}
\right) \\[4pt]
&= \left(
\begin{aligned}
& \sum_{l=1}^{3} a(l)\frac{H_z(i,j+l-1,k)-H_z(i,j-l,k)}{\Delta y} \\
& -\sum_{l=1}^{3} a(l)\frac{H_y(i,j,k+l-1)-H_y(i,j,k-l)}{\Delta z}
\end{aligned}
\right)
\end{aligned}
\tag{6.20a}
$$

$$
\begin{aligned}
&\left(
\begin{aligned}
& j\omega\varepsilon_0 \sum_{i',j',k'} E_y(i',j',k') \left[\delta_{i',i}\delta_{j'+1/2,j+1/2}\delta_{k',k} + \varepsilon_{ry}^{i',i}\varepsilon_{ry}^{j'+1/2,j+1/2}\varepsilon_{ry}^{k',k} \right] \\
& \quad + \sum_{i',j',k'} E_y(i',j',k')\sigma_{ey}^{i',i}\sigma_{ey}^{j'+1/2,j+1/2}\sigma_{ey}^{k',k}
\end{aligned}
\right) \\[4pt]
&= \left(
\begin{aligned}
& \sum_{l=1}^{3} a(l)\frac{H_x(i,j,k+l-1)-H_x(i,j,k-l)}{\Delta z} \\
& -\sum_{l=1}^{3} a(l)\frac{H_z(i+l-1,j,k)-H_z(i-l,j,k)}{\Delta x}
\end{aligned}
\right)
\end{aligned}
\tag{6.20b}
$$

$$
\begin{aligned}
&\left(
\begin{aligned}
& j\omega\varepsilon_0 \sum_{i',j',k'} E_z(i',j',k') \left[\delta_{i',i}\delta_{j',j}\delta_{k'+1/2,k+1/2} + \varepsilon_{rz}^{i',i}\varepsilon_{rz}^{j',j}\varepsilon_{rz}^{k'+1/2,k+1/2} \right] \\
& \quad + \sum_{i',j',k'} E_z(i',j',k')\sigma_{ez}^{i',i}\sigma_{ez}^{j',j}\sigma_{ez}^{k'+1/2,k+1/2}
\end{aligned}
\right) \\[4pt]
&= \left(
\begin{aligned}
& \sum_{l=1}^{3} a(l)\frac{H_y(i+l-1,j,k)-H_y(i-l,j,k)}{\Delta x} \\
& -\sum_{l=1}^{3} a(l)\frac{H_x(i,j+l-1,k)-H_x(i,j-l,k)}{\Delta y}
\end{aligned}
\right)
\end{aligned}
\tag{6.20c}
$$

$$
\begin{aligned}
&\left(
\begin{aligned}
& j\omega\mu_0 \sum_{i',j',k'} H_x(i',j',k') \left[\delta_{i',i}\delta_{j'+1/2,j+1/2}\delta_{k'+1/2,k+1/2} + \mu_{rx}^{i',i}\mu_{rx}^{j'+1/2,j+1/2}\mu_{rx}^{k'+1/2,k+1/2} \right] \\
& \quad + \sum_{i',j',k'} H_x(i',j',k')\sigma_{mx}^{i',i}\sigma_{mx}^{j'+1/2,j+1/2}\sigma_{mx}^{k'+1/2,k+1/2}
\end{aligned}
\right) \\[4pt]
&= \left(
\begin{aligned}
& \sum_{l=1}^{3} a(l)\frac{E_y(i,j,k+l)-E_y(i,j,k-l+1)}{\Delta z} \\
& -\sum_{l=1}^{3} a(l)\frac{E_z(i,j+l,k)-E_z(i,j-l+1,k)}{\Delta y}
\end{aligned}
\right)
\end{aligned}
\tag{6.20d}
$$

$$\left(\begin{array}{l} j\omega\mu_0 \sum\limits_{i',j',k'} H_y(i',j',k')\left[\delta_{i'+1/2,i+1/2}\delta_{j',j}\delta_{k'+1/2,k+1/2} + \mu_{ry}^{i'+1/2,i+1/2}\mu_{ry}^{j',j}\mu_{ry}^{k'+1/2,k+1/2}\right] \\ + \sum\limits_{i',j',k'} H_y(i',j',k')\sigma_{my}^{i'+1/2,i+1/2}\sigma_{my}^{j',j}\sigma_{my}^{k'+1/2,k+1/2} \end{array}\right)$$

$$= \left(\begin{array}{l} \sum\limits_{l=1}^{3} a(l)\frac{E_z(i+l,j,k)-E_z(i-l+1,j,k)}{\Delta x} \\ -\sum\limits_{l=1}^{3} a(l)\frac{E_x(i,j,k+l)-E_x(i,j,k-l+1)}{\Delta z} \end{array}\right) \tag{6.20e}$$

$$\left(\begin{array}{l} j\omega\mu_0 \sum\limits_{i',j',k'} H_z(i',j',k')\left[\delta_{i'+1/2,i+1/2}\delta_{j'+1/2,j+1/2}\delta_{k',k} + \mu_{rz}^{i'+1/2,i+1/2}\mu_{rz}^{j'+1/2,j+1/2}\mu_{rz}^{k',k}\right] \\ + \sum\limits_{i',j',k'} H_z(i',j',k')\sigma_{mz}^{i'+1/2,i+1/2}\sigma_{mz}^{j'+1/2,j+1/2}\sigma_{mz}^{k',k} \end{array}\right)$$

$$= \left(\begin{array}{l} \sum\limits_{l=1}^{3} a(l)\frac{E_x(i,j+l,k)-E_x(i,j-l+1,k)}{\Delta y} \\ -\sum\limits_{l=1}^{3} a(l)\frac{E_y(i+l,j,k)-E_y(i-l+1,j,k)}{\Delta x} \end{array}\right). \tag{6.20f}$$

The following coefficients are utilized in the update equations (6.20):

$$\varepsilon_{r\alpha}^{i',i} = \frac{1}{\Delta x}\int\limits_{-\infty}^{\infty} \varepsilon_{r\alpha}(x)\tilde{\phi}_{i'}(x)\phi_i(x)dx \qquad \mu_{r\alpha}^{i',i} = \frac{1}{\Delta x}\int\limits_{-\infty}^{\infty} \mu_{r\alpha}(x)\tilde{\phi}_{i'}(x)\phi_i(x)dx$$

$$\varepsilon_{r\alpha}^{j',j} = \frac{1}{\Delta y}\int\limits_{-\infty}^{\infty} \varepsilon_{r\alpha}(y)\tilde{\phi}_{j'}(y)\phi_j(y)dy \qquad \mu_{r\alpha}^{j',j} = \frac{1}{\Delta y}\int\limits_{-\infty}^{\infty} \mu_{r\alpha}(y)\tilde{\phi}_{j'}(y)\phi_j(y)dy \tag{6.21}$$

$$\varepsilon_{r\alpha}^{k',k} = \frac{1}{\Delta z}\int\limits_{-\infty}^{\infty} \varepsilon_{r\alpha}(z)\tilde{\phi}_{k'}(z)\phi_k(z)dz \qquad \mu_{r\alpha}^{k',k} = \frac{1}{\Delta z}\int\limits_{-\infty}^{\infty} \mu_{r\alpha}(z)\tilde{\phi}_{k'}(z)\phi_k(z)dz$$

and

$$\sigma_{e\alpha}^{i',i} = \frac{1}{\Delta x} \int_{-\infty}^{\infty} \sigma_\alpha^e(x) \tilde{\phi}_{i'}(x) \phi_i(x) dx \qquad \sigma_{m\alpha}^{i',i} = \frac{1}{\Delta x} \int_{-\infty}^{\infty} \sigma_\alpha^m(x) \tilde{\phi}_{i'}(x) \phi_i(x) dx$$

$$\sigma_{e\alpha}^{j',j} = \frac{1}{\Delta y} \int_{-\infty}^{\infty} \sigma_\alpha^e(y) \tilde{\phi}_{j'}(y) \phi_j(y) dy \qquad \sigma_{m\alpha}^{j',j} = \frac{1}{\Delta y} \int_{-\infty}^{\infty} \sigma_\alpha^m(y) \tilde{\phi}_{j'}(y) \phi_j(y) dy \qquad (6.22)$$

$$\sigma_{e\alpha}^{k',k} = \frac{1}{\Delta z} \int_{-\infty}^{\infty} \sigma_\alpha^e(z) \tilde{\phi}_{k'}(z) \phi_k(z) dz \qquad \sigma_{m\alpha}^{k',k} = \frac{1}{\Delta z} \int_{-\infty}^{\infty} \sigma_\alpha^m(z) \tilde{\phi}_{k'}(z) \phi_k(z) dz$$

where α is x, y, or z.

6.2   1D APPLICATION: DIELECTRIC SLAB LOADED FABRY-PEROT RESONATOR

To keep the presentation simple, a 1D problem is preferred as a numerical example. The considered 1D problem is the dielectric slab loaded Fabry-Perot resonator structure depicted in Figure 6.1. The dielectric slab is assumed to have zero magnetic and electric conductivity.

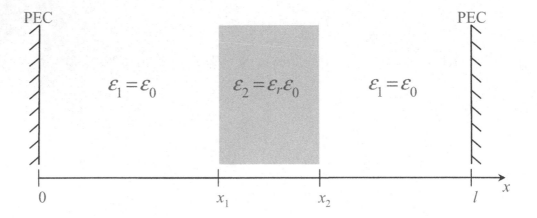

Figure 6.1: Dielectric loaded Fabry-Perot resonator.

FORMULATION

Inside the resonator, Maxwell's vector curl equations can be reduced into two scalar equations:

$$\frac{\partial H_y}{\partial x} - j\omega\varepsilon_z E_z = 0 \tag{6.23a}$$

$$\frac{\partial E_z}{\partial x} - j\omega\mu_y H_y = 0. \tag{6.23b}$$

Using the one-dimensional grid shown in Figure 4.2, the 1D MRFD update equations for (6.23) can be derived by simplifying (6.20):

$$j\omega \sum_{i'} E_z(i')\alpha_{i',i} = \sum_{l=1}^{3} a(l)\frac{\left[H_y(i+l-1) - H_y(i-l)\right]}{\Delta x} \tag{6.24a}$$

$$j\omega\mu_0 H_y(i) = \sum_{l=1}^{3} a(l)\frac{\left[E_z(i+l) - E_z(i-l+1)\right]}{\Delta x} \tag{6.24b}$$

where

$$\alpha_{i',i} = \varepsilon_0 \left[\delta_{i',i} + \varepsilon_{rz}^{i',i}\right] \tag{6.25}$$

$$\varepsilon_{rz}^{i',i} = \frac{(\varepsilon_r - 1)}{\Delta x} \int_{x_1}^{x_2} \tilde{\phi}_{i'}(x)\phi_i(x)dx \tag{6.26}$$

Some cross terms between MRFD update equations will appear around the vicinity of the discontinuity. Consider Figure 6.2 where there is an air-dielectric interface coinciding with the grid line at $x = n\Delta x$. The discretization of (6.23a) in the vicinity of the interface results in three coupled MRFD update equations:

$$j\omega E(n-1)\alpha_{n-1,n-1} + j\omega E(n)\alpha_{n-1,n} = \sum_{l=1}^{3} a(l)\frac{\left[H_y(n+l-2) - H_y(n-l-1)\right]}{\Delta x} \tag{6.27a}$$

$$j\omega E(n)\alpha_{n,n} = \sum_{l=1}^{3} a(l)\frac{\left[H_y(n+l-1) - H_y(n-l)\right]}{\Delta x} \tag{6.27b}$$

$$j\omega E(n+1)\alpha_{n+1,n+1} + j\omega E(n)\alpha_{n+1,n} = \sum_{l=1}^{3} a(l)\frac{\left[H_y(n+l) - H_y(n-l+1)\right]}{\Delta x} \tag{6.27c}$$

with the α coefficients given by:

$$\alpha_{n-1,n-1} = \varepsilon_1 \tag{6.28a}$$

$$\alpha_{n,n} = \frac{\varepsilon_1 + \varepsilon_2}{2} \tag{6.28b}$$

$$\alpha_{n+1,n+1} = \varepsilon_2 \tag{6.28c}$$

$$\alpha_{n-1,n} = 0.00037(\varepsilon_1 - \varepsilon_2) \tag{6.28d}$$

$$\alpha_{n+1,n} = 0.00037(\varepsilon_2 - \varepsilon_1). \tag{6.28e}$$

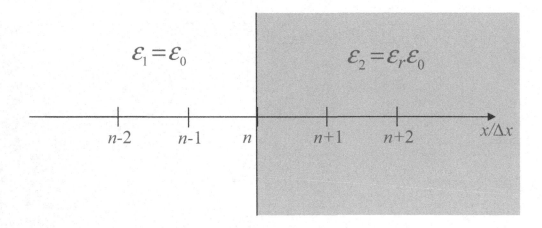

Figure 6.2: Interface between two different dielectric mediums.

Notice that ignoring the $\alpha_{n-1,n}$ and $\alpha_{n+1,n}$ coefficients leads to the homogeneous MRFD formulation with the averaging of material parameters.

NUMERICAL RESULTS

The dielectric loaded Fabry-Perot resonator depicted in Figure 6.1 with $l = 1$ m, $x_1 = 0.4$ m, and $x_2 = 0.6$ m is analyzed for two different dielectric constants. The first two resonance frequencies of the resonator are calculated with both homogeneous (with pointwise sampling of material parameters and averaging of material parameters at interfaces) and inhomogeneous MRFD formulations with a cell size of $\Delta x = 4$ cm, and the results are compared to the analytical values in Table 6.1.

Results show insignificant differences between the two formulations even if the contrast ratio of the dielectric constants of two neighboring mediums is a high value like 64:1. This is simply a result of the two formulations being identical except for the non-diagonal α coefficients ($\alpha_{n-1,n}$ and $\alpha_{n+1,n}$), which are very small compared to the diagonal α coefficients.

In conclusion, the homogeneous MRFD formulation based on the CDF(2,2) wavelet, utilizing the use of pointwise sampling and averaging of material parameters, is found to be accurate enough to replace the inhomogeneous MRFD formulation.

Table 6.1: Resonance frequencies of the dielectric loaded resonator

Dielectric Constant	Analytic	Inhomog. MRFD	Homog. MRFD
16	55.306 MHz	55.426 MHz	55.428 MHz
	187.372 MHz	187.111 MHz	187.108 MHz
64	28.382 MHz	28.457 MHz	28.458 MHz
	100.326 MHz	100.586 MHz	100.591 MHz

CHAPTER 7

Conclusion

This chapter concludes the book. The contributions of the work are summarized and future work to further improve the MRFD technique is suggested.

The primary focus of this research was to develop new frequency domain computational electromagnetic techniques based on multiresolution analysis. It was anticipated that the new formulations exhibit superior efficiency characteristics compared to the finite difference frequency domain scheme. The new general formulation, named the multiresolution frequency domain technique, is based on Battle-Lemarie and biorthogonal CDF scaling functions.

The memory and simulation time requirements of various MRFD schemes based on different wavelet bases are compared by calculating the resonance frequency of homogeneous and inhomogeneous one-dimensional resonators. It was concluded that the CDF(2,2) wavelet-based MRFD scheme exhibits the highest efficiency among other schemes based on CDF wavelets and Battle-Lemarie wavelets.

The CDF(2,2) wavelet-based MRFD formulation is also implemented for two and three-dimensional closed space problems. The propagation characteristics of various waveguiding structures are analyzed and the resonance frequencies of three-dimensional resonators are computed. In a one-dimensional case, the MRFD method showed significant savings in terms of computer memory and simulation time compared to the FDFD method; however extension to two and three-dimensional problems produced even more pronounced savings.

In order to emphasize the versatility of the MRFD technique, scattering problems are also characterized with the new formulation. An unsplit scattering field formulation is developed to model open problems, and the perfectly matched layer (PML) technique is employed in order to truncate the computational space. Similar to the closed space problems, MRFD technique produced improved efficiency figures.

Finally, a new multiresolution technique is proposed in order to model inhomogeneous mediums in a mathematically correct way, which is called the inhomogeneous MRFD formulation. This formulation is compared to the simple MRFD formulation to conclude that the simple MRFD formulation provides enough accuracy even for inhomogeneous problems.

The current research can be extended in many directions. The following suggestions are only a few:

The introduced MRFD formulations are based on only the scaling functions. The MRFD algorithm based on both scaling and wavelet functions promises a multigrid formulation which can be used to introduce finer resolutions around discontinuities. This could especially be used to model very thin layers or sub-cell size discontinuities.

Finite difference or multiresolution modeling of PEC objects that do not conform to the rectangular grid has proven to be a challenge. The MRFD method can be extended to model non-conformal PEC-dielectric interfaces.

Extending the formulation to model complex mediums such as chiral or gyrotropic mediums can be another research direction.

APPENDIX A

Resonance Frequencies of a Dielectric Loaded Fabry-Perot Resonator

The Fabry-Perot resonator with a dielectric slab inserted between its plates is shown in Figure A.1. If plane wave propagation and lossless medium inside the plates is assumed, the propagation constant β can be calculated from:

$$k \cot \beta_1 S_1 \left(\cot \beta S_2 + \cot \beta x_1\right) + \cot \beta S_2 \cot \beta x_1 - k^2 = 0 \qquad (A.1)$$

where β is the propagation constant inside the air regions, β_1 is the propagation constant inside the dielectric slab, and

$$k = \sqrt{\frac{\mu_1/\varepsilon_1}{\mu_2/\varepsilon_2}} = \sqrt{\frac{\mu_1 \varepsilon_2}{\mu_2 \varepsilon_1}} . \qquad (A.2)$$

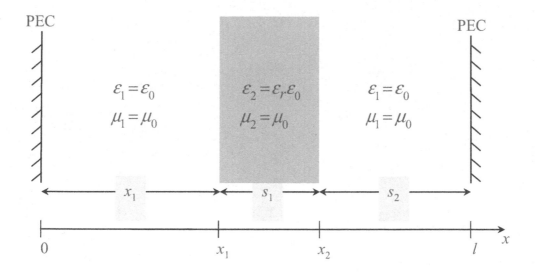

Figure A.1: Dielectric loaded Fabry-Perot resonator.

If a non-magnetic dielectric slab, as shown in Figure A.1, is assumed, k and β_1 can be simplified as:

$$k = \sqrt{\varepsilon_r} \tag{A.3a}$$

$$\beta_1 = \sqrt{\varepsilon_r}\beta . \tag{A.3b}$$

Substituting (A.3) into (A.1) yields:

$$\sqrt{\varepsilon_r} \cot\left(\sqrt{\varepsilon_r}\beta S_1\right)\left(\cot\beta S_2 + \cot\beta x_1\right) + \cot\beta S_2 \cot\beta x_1 - \varepsilon_r = 0 . \tag{A.4}$$

(A.4) can be solved to calculate β. Note that (A.4) is a transcendental equation and many values of β, each corresponding to a resonant mode, can satisfy it.

Once β values are obtained, the corresponding resonance frequencies can be calculated by:

$$f = \frac{\beta}{2\pi\sqrt{\varepsilon_0\mu_0}} . \tag{A.5}$$

APPENDIX B

Propagation Constant of Rectangular Waveguide Structures

The uniform rectangular waveguide structure is shown in Figure B.1a. The propagation constants of the $TE_{Z,m,n}$ modes of this structure can be calculated by:

$$\beta_{m,n} = 2\pi \sqrt{\mu\varepsilon \left(f^2 - f_{C,m,n}^2 \right)} \tag{B.1}$$

where f_C is the cut-off frequency given by

$$f_{C,m,n} = \frac{1}{\sqrt{\mu\varepsilon}} \sqrt{\left(\frac{m}{a}\right)^2 + \left(\frac{n}{b}\right)^2} \, . \tag{B.2}$$

The propagation constants of the $TM_{X,m,n}$ modes of the partially filled rectangular waveguide structure (Figure B.1b) can be calculated by:

$$\frac{k_{x1}}{\varepsilon_1} \tan k_{x1}d = -\frac{k_{x2}}{\varepsilon_2} \tan \left[k_{x2}(a - h)\right] \tag{B.3}$$

where

$$k_{x1} = \sqrt{\omega^2\varepsilon_1\mu_1 - \left(\frac{n\pi}{b}\right)^2 - \beta^2} \tag{B.4a}$$

$$k_{x2} = \sqrt{\omega^2\varepsilon_2\mu_2 - \left(\frac{n\pi}{b}\right)^2 - \beta^2} \tag{B.4b}$$

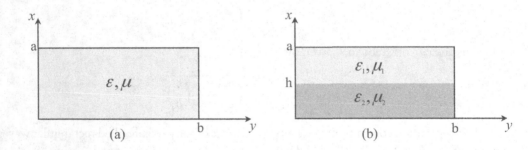

Figure B.1: Rectangular waveguide structures.

APPENDIX C

Resonance Frequencies of Rectangular Cavity Resonators

The resonance frequencies of the uniform cavity resonator shown in Figure C.1a are given by:

$$f_{r,m,n,p} = \frac{1}{2\sqrt{\varepsilon\mu}}\sqrt{\left(\frac{m}{a}\right)^2 + \left(\frac{n}{b}\right)^2 + \left(\frac{p}{c}\right)^2} \tag{C.1}$$

The m, n, p coefficients for the $TE_{m,n,p}$ modes are

$$
\begin{aligned}
m &= 0, 1, 2, \dots \\
n &= 0, 1, 2, \dots \\
p &= 1, 2, 3, \dots \\
m &= n = 0 \quad \text{excepted}
\end{aligned}
\tag{C.2}
$$

and the m, n, p coefficients for the $TM_{m,n,p}$ modes are

$$
\begin{aligned}
m &= 1, 2, 3, \dots \\
n &= 1, 2, 3, \dots \\
p &= 0, 1, 2, \dots
\end{aligned}
\tag{C.3}
$$

The resonance frequencies of the partially filled cavity resonator shown in Figure C.1b can be calculated by:

$$\frac{k_{x1}}{\varepsilon_1}\tan k_{x1}d = -\frac{k_{x2}}{\varepsilon_2}\tan\left[k_{x2}(a-h)\right] \tag{C.4}$$

$$\frac{k_{x1}}{\mu_1}\cot k_{x1}d = -\frac{k_{x2}}{\mu_2}\cot\left[k_{x2}(a-h)\right] \tag{C.5}$$

where

$$k_{x1} = \sqrt{\omega^2\varepsilon_1\mu_1 - \left(\frac{n\pi}{b}\right)^2 - \left(\frac{p\pi}{c}\right)^2} \tag{C.6a}$$

$$k_{x2} = \sqrt{\omega^2\varepsilon_2\mu_2 - \left(\frac{n\pi}{b}\right)^2 - \left(\frac{p\pi}{c}\right)^2} \tag{C.6b}$$

and

$$
\begin{aligned}
&n = 0, 1, 2, \ldots \\
&p = 0, 1, 2, \ldots \\
&n = p = 0 \quad \text{excepted}
\end{aligned}
\tag{C.7}
$$

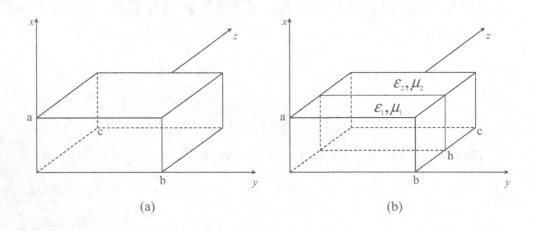

(a)   (b)

Figure C.1: Rectangular cavity resonators.

APPENDIX D

Near to Far Field Transformation

Scattering problems usually deal with far field electromagnetic responses such as bistatic radar cross sections or radiation patterns. Therefore, the fields at far distances need to be calculated. For this calculation, extending the computational domain to far distances is impractical because the required computer resources will increase rapidly as the computational domain is physically enlarged. A very attractive alternative is to calculate the far fields from the near fields by the use of the surface equivalence theorem [68].

For electromagnetic field problems, the uniqueness theorem states that when the sources and the tangential electric or magnetic fields are specified over the whole boundary surface of a given region, then the solution within this region is unique. Based on the uniqueness theorem, if the tangential electric and magnetic fields are completely known over a closed surface that bounds it, the fields in a source-free region can be uniquely determined. So, a problem can be replaced by another problem if the tangential electric and magnetic fields are the same on the boundaries of both problems. As shown in Figure D.1, the fields in the outer regions (regions of interest) of these two schematics are the same since the electric ($\bar{J}_S$) and magnetic ($\overline{M}_S$) currents on the surface of the object in the second problem will produce the tangential electric and magnetic fields which are equivalent to the fields on the boundaries of the first problem.

The equivalence theorem described above is used to calculate far fields in the MRFD method. First the scatterer is bounded by an imaginary closed contour in the computational space. Since the MRFD technique employs a Cartesian coordinate system, the closed contour is assumed to be a box for the sake of convenience. Secondly, the electric and magnetic currents ($\bar{J}$ and $\bar{M}$) on the imaginary surface are calculated from the scattered magnetic and electric fields tangent to this surface. The currents $\bar{J}$ and $\bar{M}$ produce the scattered fields only outside of the imaginary surface.

For the scattered field formulation we can write

$$\bar{J}_{total} = \hat{n} \times \bar{H}_{total} = \hat{n} \times \left(\bar{H}_{inc} + \bar{H}_{scat}\right) = \bar{J}_{inc} + \bar{J} \tag{D.1a}$$

$$\bar{M}_{total} = -\hat{n} \times \bar{E}_{total} = -\hat{n} \times \left(\bar{E}_{inc} + \bar{E}_{scat}\right) = \bar{M}_{inc} + \bar{M} \tag{D.1b}$$

The currents $\bar{J}_{inc}$ and $\bar{M}_{inc}$ produce zero field everywhere outside of the imaginary surface, hence they are not used in the scattered field formulation.

If the internal space of the imaginary surface is assumed to have zero electric and magnetic fields in the absence of the source of the incident fields, then the fictitious currents will be $\bar{J}_s =$

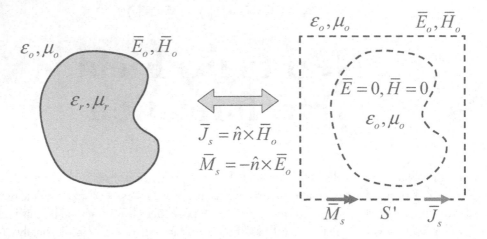

Figure D.1: Two-dimensional equivalent problems in the outer regions.

$\hat{n} \times \bar{H}_o$ and $\bar{M}_s = -\hat{n} \times \bar{E}_o$ where $\bar{H}_o$ and $\bar{E}_o$ are the scattered fields on the imaginary surface in the grid and $\hat{n}$ is the unit normal vector that points outward from the surface. These fictitious currents were then transformed to the far field using vector potentials:

$$E_r = 0 \tag{D.2a}$$

$$E_\theta = -\frac{jke^{-jkr}}{4\pi r}\left(L_\phi + \eta_o N_\theta\right) \tag{D.2b}$$

$$E_\phi = +\frac{jke^{-jkr}}{4\pi r}\left(L_\theta - \eta_o N_\phi\right) \tag{D.2c}$$

$$H_r = 0 \tag{D.2d}$$

$$H_\theta = +\frac{jke^{-jkr}}{4\pi r}\left(N_\phi - \frac{L_\theta}{\eta_o}\right) \tag{D.2e}$$

$$H_\phi = -\frac{jke^{-jkr}}{4\pi r}\left(\bar{N}_\theta + \frac{L_\phi}{\eta_o}\right) \tag{D.2f}$$

where (r, θ, ϕ) is the location of the observation point in spherical coordinates, $\eta_o = \sqrt{\mu_o/\varepsilon_o}$ is the intrinsic impedance of free space, and

$$N_\theta = \iint_s \left(J_x \cos\theta \cos\phi + J_y \cos\theta \sin\phi - J_z \sin\theta\right) e^{+jkr'\cos\psi} ds' \tag{D.3a}$$

$$N_\phi = \iint_S \left(-J_x \sin\phi + J_y \cos\phi \right) e^{+jkr' \cos\psi} ds' \tag{D.3b}$$

$$L_\theta = \iint_S \left(M_x \cos\theta \cos\phi + M_y \cos\theta \sin\phi - M_z \sin\theta \right) e^{+jkr' \cos\psi} ds' \tag{D.3c}$$

$$L_\phi = \iint_S \left(-M_x \sin\phi + M_y \cos\phi \right) e^{+jkr' \cos\psi} ds' . \tag{D.3d}$$

Here, r is the magnitude of the observation point vector from the origin in the vicinity of the scatterer, r' is the magnitude of the source point vector, and ψ is the angle between $\vec{r}$ and $\vec{r}'$ as shown in Fig. D.2.

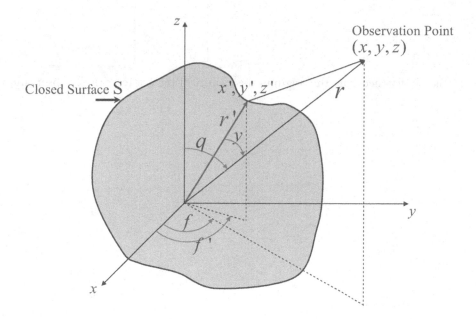

Figure D.2: Far field observation point and source point.

In many scattering problems, calculation of the radar cross section is the main goal. For a θ-polarized incident plane wave, the co-polarized and cross-polarized bistatic radar cross sections, $\sigma_{\theta\theta}$ and $\sigma_{\phi\theta}$, are defined as

$$\sigma_{\theta\theta} = \lim_{r \to \infty} 4\pi r^2 \frac{|E_\theta^s|^2}{|E_\theta^i|^2} \tag{D.4a}$$

$$\sigma_{\phi\theta} = \lim_{r \to \infty} 4\pi r^2 \frac{|E_\phi^s|^2}{|E_\theta^i|^2} \tag{D.4b}$$

where E^s and E^i are scattered and incident electric fields, respectively.

Similarly, for a two-dimensional case, the co-polarized and cross-polarized bistatic echo widths can be written as:

$$SW\sigma_{zz} = \lim_{\rho \to \infty} \left[2\pi\rho \frac{|E_z^s|^2}{|E_z^i|^2} \right] \tag{D.5a}$$

$$SW\sigma_{\phi z} = \lim_{\rho \to \infty} \left[2\pi\rho \frac{|E_\phi^s|^2}{|E_z^i|^2} \right] \tag{D.5b}$$

where ρ is the distance from the origin in the vicinity of the scatterer to the observation point.

Bibliography

[1] E. Chiprout and M. S. Nakhla, *Asymptotic Waveform Equation and Moment Matching for Interconnect Analysis*, Norwell, MA: Kluwer, 1994. Cited on page(s) 1

[2] K. Sabetfakhri and L. P. B. Katehi, "Analysis of integrated millimeter-wave and submillimeter-wave waveguides using orthonormal wavelet expansions," *IEEE Transactions on Microwave Theory and Techniques*, vol. 42, no. 12, part 2, pp. 2412–2422, December 1994. DOI: 10.1109/22.339775 Cited on page(s) 2

[3] B. Z. Steinberg and Y. Leviatan, "On the use of wavelet expansions in the method of moments," *IEEE Transactions on Antennas and Propagation*, vol. 41, no. 5, pp. 610–619, May 1993. DOI: 10.1109/8.222280 Cited on page(s) 2

[4] R. L. Wagner and W. C. Chew, "Study of wavelets for the solution of electromagnetic integral equations," *IEEE Transactions on Antennas and Propagation*, vol. 43, no. 8, pp. 802–810, August 1995. DOI: 10.1109/8.402199 Cited on page(s) 2

[5] I. Barba, J. Represa, M. Fujii, and W. J. R. Hoefer, "Multi-resolution 2D-TLM technique using Haar wavelets," *IEEE MTT-S International Microwave Symposium Digest*, vol. 1, pp. 243–246, Boston, MA, USA, June 2000. DOI: 10.1109/MWSYM.2000.860962 Cited on page(s) 2

[6] T. Dogaru and L. Carin, "Multiresolution time-domain using CDF biorthogonal wavelets," *IEEE Transactions on Microwave Theory and Techniques*, vol. 49, no. 5, pp. 902–912, May 2001. DOI: 10.1109/22.920147 Cited on page(s) 2, 19, 22, 29

[7] M. Fujii and W. J. R. Hoefer, "A three-dimensional Haar-wavelet-based multiresolution analysis similar to the FDTD method–derivation and application," *IEEE Transactions on Microwave Theory and Techniques*, vol. 46, no. 12, part 2, pp. 2463–2475, December 1998. DOI: 10.1109/22.739236 Cited on page(s) 2, 19

[8] M. Fujii and W. J. R. Hoefer, "A wavelet formulation of the finite-difference method: Full-vector analysis of optical waveguide junctions," *IEEE Journal of Quantum Electronics*, vol. 37, no. 8, pp. 1015–1029, August 2001. DOI: 10.1109/3.937391 Cited on page(s) 2, 19

[9] M. Krumpholz and L. P. B. Katehi, "MRTD: new time-domain schemes based on multiresolution analysis," *IEEE Transactions on Microwave Theory and Techniques*, vol. 44, no. 4, pp. 555–571, April 1996. DOI: 10.1109/22.491023 Cited on page(s) 2, 19, 35

[10] R. F. Harrington, *Field Computation by Moment Methods*. Malabar, FL: Krieger, 1982. Cited on page(s) 5, 19

[11] R. W. Clough, "The finite element in plane stress analysis," *2nd ASCE Conference on Electronic Computation*, Pittsburgh, PA, 1960. Cited on page(s) 5

[12] P. B. Johns and R. L. Beurie, "Numerical solution of 2-dimensional scattering problems using a transmission-line matrix," *Proceedings of the Institution of Electrical Engineers*, vol. 118, no. 9, pp. 1203–1208, 1971. DOI: 10.1049/piee.1971.0217 Cited on page(s) 5

[13] A. Taflove and S. C. Hagness, *Computational Electrodynamics: The Finite-Difference Time-Domain Method, 3rd ed.*, Boston, MA: Artech House, 2005. Cited on page(s) 5, 63

[14] M. Al Sharkawy, V. Demir, and A. Z. Elsherbeni, "Efficient analysis of electromagnetic scattering problems using a parallel-multigrid iterative multi-region algorithm," *Antennas and Propagation Society International Symposium*, pp. 3985–3988, Albuquerque, NM, USA, 2006. DOI: 10.1109/APS.2006.1711499 Cited on page(s) 5, 6

[15] L. Kuzu, V. Demir, A. Z. Elsherbeni, and E. Arvas, "Electromagnetic scattering from arbitrarily shaped three-dimensional chiral objects using the finite difference frequency domain method," *URSI National Radio Science Meeting*, p. 333, Albuquerque, NM, USA, 2006. DOI: 10.2528/PIER06083104 Cited on page(s) 5

[16] M. Salazar-Palma, T. K. Sarkar, L. E. Garcia-Castillo, T. Roy, and A. Djordjevic, *Iterative and Self Adaptive Finite-Elements in Electromagnetic Modeling*. Boston, MA: Artech House, 1998. Cited on page(s) 5

[17] A. Thom and C. J. Apelt, *Field Computations in Engineering and Physics*. London: D. Van Nostrand, 1961. Cited on page(s) 5

[18] K. S. Yee, "Numerical solution of initial boundary value problems involving Maxwell's equations in isotropic media," *IEEE Transactions on Antennas and Propagation*, vol. 14, pp. 302–307, 1966. DOI: 10.1109/TAP.1966.1138693 Cited on page(s) 5, 8

[19] J. I. Glaser, "Numerical solution of waveguide scattering problems by finite-difference Green's functions," *IEEE Transactions on Microwave Theory and Techniques*, vol. 18, no. 8, pp. 436–443, August 1970. DOI: 10.1109/TMTT.1970.1127265 Cited on page(s) 6

[20] M. J. Hagmann, O. P. Gandhi, and C. H. Durney, "Numerical calculation of electromagnetic energy deposition for a realistic model of man," *IEEE Transactions on Microwave Theory and Techniques*, vol. 27, no. 9, pp. 804–809, September 1979. DOI: 10.1109/TMTT.1979.1129735 Cited on page(s) 6

[21] B. J. McCartin and J. F. Dicello, "Three dimensional finite difference frequency domain scattering computation using the Control Region Approximation," *IEEE Transactions on Magnetics*, vol. 25, no. 4, pp. 3092–3094, July 1989. DOI: 10.1109/20.34378 Cited on page(s) 6

[22] L. Kuzu, "Electromagnetic scattering from chiral materials using the FDFD method," Ph.D. Dissertation, Department of Electrical Engineering, Syracuse University, Syracuse, NY, May 2006. Cited on page(s) 6

[23] M. N. O. Sadiku, *Numerical techniques in electromagnetics, 2nd ed.*, Boca Raton, FL: CRC Press, 2000. DOI: 10.1201/9781420058277 Cited on page(s) 6

[24] I. Daubechies, *Ten Lectures on Wavelets*. Philadelphia, PA: Society for Industrial and Applied Mathematics, 1992. DOI: 10.1137/1.9781611970104 Cited on page(s) 17, 22

[25] M. Krumpholz, C. Huber, and P. Russer, "A field theoretical comparison of FDTD and TLM," *IEEE Transactions on Microwave Theory and Techniques*, vol. 43, no. 8, pp. 1935–1950, August 1995. DOI: 10.1109/22.402284 Cited on page(s) 19

[26] C. D. Sarris, L. P. B. Katehi, and J. F. Harvey, "Application of multiresolution analysis to the modeling of microwave and optical structures," *Optical and Quantum Electronics*, vol. 32, no. 6–8, pp. 657-679, August 2000. DOI: 10.1023/A:1007040007716 Cited on page(s) 19, 20

[27] G. W. Pan, *Wavelets in Electromagnetics and Device Modeling*. Hoboken, New Jersey: John Wiley & Sons Inc., 2003. DOI: 10.1002/0471433918 Cited on page(s) 22

[28] X. Zhu, T. Dogaru, and L. Carin, "Analysis of the CDF biorthogonal MRTD method with application to PEC targets," *IEEE Transactions on Microwave Theory and Techniques*, vol. 51, no. 9, pp. 2015–2022, September 2003. DOI: 10.1109/TMTT.2003.815874 Cited on page(s) 22, 31

[29] P. G. Lemarie, "Ondelettes a localization exponentielle," *Journal de Mathematiques Pures et Appliquees*, vol. 67, pp. 227–236, 1988. Cited on page(s) 22

[30] G. Battle, "A block spin construction of ondelettes," *Communications in Mathematical Physics*, vol. 110, pp. 601–615, 1987. DOI: 10.1007/BF01205550 Cited on page(s) 22

[31] A. Cohen, I. Daubechies, and J.-C. Feauveau, "Biorthogonal bases of compactly supported wavelets," *Communications on Pure and Applied Mathematics*, vol. 45, no. 5, pp. 485–560, 1992. DOI: 10.1002/cpa.3160450502 Cited on page(s) 22

[32] X. Zhu, L. Carin, and T. Dogaru, "Parallel implementation of the biorthogonal multiresolution time-domain method," *Journal of the Optical Society of America A: Optics, Image Science, and Vision*, vol. 20, no. 5, pp. 844–855, May 2003. DOI: 10.1364/JOSAA.20.000844 Cited on page(s) 22

[33] X. Zhu, T. Dogaru, and L. Carin, "Three-dimensional biorthogonal multiresolution time-domain method and its application to electromagnetic scattering problems," *IEEE Transactions on Antennas and Propagation*, vol. 51, no. 5, pp. 1085–1092, May 2003. DOI: 10.1109/TAP.2003.811527 Cited on page(s) 22

[34] K. L. Shlager and J. B. Schneider, "A 2-D dispersion analysis of the W-MRTD method using CDF biorthogonal wavelets," *IEEE Antennas and Propagation Society International Symposium*, vol. 3, pp. 244–247, San Antonio, TX, 2002. DOI: 10.1109/APS.2002.1018201 Cited on page(s) 22

[35] T. Dogaru and L. Carin, "Efficient time-domain electromagnetic analysis using CDF biorthogonal wavelet expansion," *IEEE MTT-S International Microwave Symposium Digest*, vol. 3, pp. 2019–2022, Phoenix, AZ, 2001. DOI: 10.1109/MWSYM.2001.967307 Cited on page(s) 28

[36] T. Dogaru and L. Carin, "Scattering analysis by the multiresolution time-domain method using compactly supported wavelet systems," *IEEE Transactions on Microwave Theory and Techniques*, vol. 50, no. 7, pp. 1752–1760, July 2002. DOI: 10.1109/TMTT.2002.800426 Cited on page(s) 29

[37] C. A. Balanis, "Measurements of dielectric constants and loss tangents at E-Band using a Fabry-Perot interferometer," NASA D-5583, December 1969 Cited on page(s) 38

[38] J. A. Pereda, A. Vegas, L. F. Velarde, and O. Gonzalez, "An FDFD eigenvalue formulation for computing port solutions in FDTD simulators," *Microwave and Optical Technology Letters*, vol. 45, no. 1, pp. 1–3, April 2005. DOI: 10.1002/mop.20704 Cited on page(s) 40

[39] D. H. Choi and W. J. R. Hoefer, "The finite difference-time domain method and its application to eigenvalue problems," *IEEE Transactions on Microwave Theory and Techniques*, vol. 34, no. 12, pp. 1464–1470, December 1986. DOI: 10.1109/TMTT.1986.1133564 Cited on page(s) 40

[40] L. L. Liou, M. Mah, and J. Cook, "An equivalent circuit approach for microstrip component analysis using the FDTD method," *IEEE Microwave and Guided Wave Letters*, vol. 8, no. 10, pp. 330–332, October 1998. DOI: 10.1109/75.735411 Cited on page(s) 40

[41] A. Asi and L. Shafai, "Dispersion analysis of anisotropic inhomogeneous waveguides using compact 2D-FDTD," *Electronics Letters*, vol. 28, no. 15, pp. 1451–1452, July 1992. DOI: 10.1049/el:19920923 Cited on page(s) 40

[42] S. Xiao and R. Vahldieck, "An efficient 2-D FDTD algorithm using real variables," *IEEE Microwave and Guided Wave Letters*, vol. 3, no. 5, pp. 127–129, May 1993. DOI: 10.1109/75.217204 Cited on page(s) 40

[43] S. Xiao, R. Vahldieck, and H. Jin, "Full-wave analysis of guided wave structures using a novel 2-D FDTD," *IEEE Microwave and Guided Wave Letters*, vol. 2, no. 5, pp. 165–167, May 1992. DOI: 10.1109/75.134342 Cited on page(s) 40

[44] A. Asi and L. Shafai, "Multiple mode analysis of waveguides using compact FDTD," *IEEE Antennas and Propagation Society International Symposium Digest*, pp. 360–363, Ann Arbor, MI, USA, June 1993. DOI: 10.1109/APS.1993.385331 Cited on page(s) 40

[45] F. Liu, J. E. Schutt-Aine, and J. Chen, "Full-wave analysis and modeling of multiconductor transmission lines via 2-D-FDTD and signal-processing techniques," *IEEE Transactions on Microwave Theory and Techniques*, vol. 50, no. 2, pp. 570–577, February 2002. DOI: 10.1109/22.982237 Cited on page(s) 40

[46] A. C. Cangellaris, "Numerical stability and numerical dispersion of a compact 2-D/FDTD method used for the dispersion analysis of waveguides," *IEEE Microwave and Guided Wave Letters*, vol. 3, no. 1, pp. 3–5, January 1993. DOI: 10.1109/75.180672 Cited on page(s) 40

[47] M. L. Lui and Z. Chen, "A direct computation of propagation constant using compact 2-D full-wave eigen-based finite-difference frequency-domain technique," *International Conference on Computational Electromagnetics and its Applications*, pp. 78–81, Beijing, China, 1999. DOI: 10.1109/ICCEA.1999.825072 Cited on page(s) 40, 41

[48] J. N. Hwang, "A compact 2-D FDFD method for modeling microstrip structures with nonuniform grids and perfectly matched layer," *IEEE Transactions on Microwave Theory and Techniques*, vol. 53, no. 2, pp. 653–659, February 2005. DOI: 10.1109/TMTT.2004.840569 Cited on page(s) 40

[49] Y. J. Zhao, K.L. Wu, and K.K. M. Cheng, "A compact 2-D full-wave finite-difference frequency-domain method for general guided wave structures," *IEEE Transactions on Microwave Theory and Techniques*, vol. 50, no. 7, pp. 1844–1848, July 2002. DOI: 10.1109/TMTT.2002.800447 Cited on page(s) 40

[50] B. Z. Wang, X. Wang, and W. Shao, "2D full-wave finite-difference frequency-domain method for lossy metal waveguide," *Microwave and Optical Technology Letters*, vol. 42, no. 2, pp. 158–161, Julyy 2004. DOI: 10.1002/mop.20238 Cited on page(s) 40

[51] L. Y. Li and J.F. Mao, "An improved compact 2-D finite-difference frequency-domain method for guided wave structures," *IEEE Microwave and Wireless Components Letters*, vol. 13, no. 12, pp. 520–522, December 2003. DOI: 10.1109/LMWC.2003.819956 Cited on page(s) 40

[52] Q. Cao, Y. Chen, and R. Mittra, "Multiple image technique (MIT) and anisotropic perfectly matched layer (APML) in implementation of MRTD scheme for boundary truncations of microwave structures," *IEEE Transactions on Microwave Theory and Techniques*, vol. 50, no. 6, pp. 1578–1589, June 2002. DOI: 10.1109/TMTT.2002.1006420 Cited on page(s) 48

[53] R. F. Harrington, *Time-Harmonic Electromagnetic Fields*, York, PA: McGraw-Hill Book Company, Inc., 1961. Cited on page(s) 49, 59

[54] R. B. Lehoucq and D. C. Sorensen, "Deflation Techniques for an Implicitly Re-Started Arnoldi Iteration," *SIAM Journal of Matrix Analysis and Applications*, vol. 17, pp. 789–821, 1996. DOI: 10.1137/S0895479895281484 Cited on page(s) 59

[55] E. Anderson, Z. Bai, C. Bischof, S. Blackford, J. Demmel, J. Dongarra, J. D. Croz, A. Greenbaum, S. Hammarling, A. McKenney, and D. Sorensen, *LAPACK User's Guide, 3rd ed.,* Philadelphia, PA: SIAM, 1999. DOI: 10.1137/1.9780898719604 Cited on page(s) 59

[56] K. S. Kunz and R. J. Luebbers, *The Finite Difference Time Domain Method for Electromagnetics*, Boca Raton: CRC Press, 1993. Cited on page(s) 63

[57] C. M. Rappaport and L. J. Bahrmasel, "An absorbing boundary condition based on anechoic absorber for EM scattering computation," *Journal of Electromagnetic Waves and Applications*, vol. 6, no. 12, pp. 1621–1633, December 1992. DOI: 10.1163/156939392X00760 Cited on page(s) 66

[58] C. M. Rappaport, "Preliminary FDTD results from the anechoic absorber absorbing boundary condition," *IEEE Antennas and Propagation Society International Symposium*, vol. 1, pp. 544–547, July 1992. DOI: 10.1109/APS.1992.221880 Cited on page(s) 66

[59] G. Mur, "Absorbing boundary conditions for the finite-difference approximation of the time-domain electromagnetic-field equations," *IEEE Transactions on Electromagnetic Compatibility*, vol. 23, pp. 377–382, 1981. DOI: 10.1109/TEMC.1981.303970 Cited on page(s) 66

[60] Z. P. Liao, H. L. Wong, B. Yang, and Y. Yuan, "A transmitting boundary for transient wave analysis," *Scientia Sinica Series A, Mathematical Physical Astronomical & Technical Sciences*, vol. 27, pp. 1063–1076, 1984. Cited on page(s) 66

[61] J. P. Berenger, "A perfectly matched layer for the absorption of electromagnetic waves," *Journal of Computational Physics*, vol. 114, pp. 185–200, 1994. DOI: 10.1006/jcph.1994.1159 Cited on page(s) 66

[62] R. Mittra and U. Pekel, "A new look at the perfectly matched layer (PML) concept for the reflectionless absorption of electromagnetic waves," *IEEE Microwave and Guided Wave Letters*, vol. 5, no. 3, pp. 84–86, March 1995. DOI: 10.1109/75.366461 Cited on page(s) 66

[63] J. P. Berenger, "Three-dimensional perfectly matched layer for the absorption of electromagnetic waves," *Journal of Computational Physics*, vol. 127, pp. 363–379, 1996. DOI: 10.1006/jcph.1996.0181 Cited on page(s) 66

[64] D. S. Katz, E. T. Thiele, and A. Taflove, "Validation and extension to three dimensions of the Berenger PML absorbing boundary condition for FD-TD meshes," *IEEE Microwave and Guided Wave Letters*, vol. 4, no. 8, pp. 268–270, August 1994. DOI: 10.1109/75.311494 Cited on page(s) 67

[65] Z. S. Sacks, D. M. Kingsland, R. Lee, and J. F. Lee, "A perfectly matched anisotropic absorber for use as an absorbing boundary condition," *IEEE Transactions on Antennas and Propagation*, vol. 43, no. 12, pp. 1460–1463, December 1995. DOI: 10.1109/8.477075 Cited on page(s) 68

[66] Z. Wang, "*A study of the numerical solution of two dimensional electromagnetic scattering problems via the finite difference method with a perfectly matched layer boundary condition*," Master Thesis, Engineering Science, University of Mississippi, University, MS, August 1995. Cited on page(s)

[67] Y. Zhu, "Application of multi-resolution time domain (MRTD) method to electromagnetic engineering," Ph.D. Dissertation, Department of Electrical Engineering, University of South Carolina, Columbia, SC, 2004. Cited on page(s) 89

[68] C. A. Balanis, *Advanced Engineering Electromagnetics*. New York: Wiley, 1989. Cited on page(s) 82, 111

Authors' Biographies

MESUT GOKTEN

Mesut Gokten was born in Tokat, Turkey, in 1975. He graduated from the Istanbul Technical University Electronics and Communications Engineering Department in 1997. He received his M.S. (with honors) and Ph.D. degrees from the Syracuse University Electrical Engineering Department in 2003 and 2007, respectively, and served as a teaching and a research assistant from August 2000 to May 2005 at the same department. He was with Dielectric Laboratories, Inc., of Cazenovia, NY, from May 2005 to August 2006. He currently works at the R&D and Satellite Department at Turksat AS, Ankara, Turkey. His research is focused on computational electromagnetics, design, and implementation of microwave devices and systems for satellite communications.

ATEF ELSHERBENI

Dr. Atef Elsherbeni is a Professor of Electrical Engineering and Associate Dean of Engineering for Research and Graduate Programs, and the Director of the Center for Applied Electromagnetic Systems Research (CAESR) at the University of Mississippi. In 2004 he was appointed as an adjunct professor, at the Department of Electrical Engineering and Computer Science of the L.C. Smith College of Engineering and Computer Science at Syracuse University. In 2009 he was selected as Finland Distinguished Professor by the Academy of Finland and TEKES. Dr. Elsherbeni has conducted research dealing with scattering and diffraction by dielectric and metal objects, finite difference time domain analysis of passive and active microwave devices including planar transmission lines, field visualization and software development for EM education, interactions of electromagnetic waves with the human body, RFID and sensors development for monitoring soil moisture, airports noise levels, and air quality (including haze and humidity), reflector and printed antennas and antenna arrays for radars, UAV, and personal communication systems, antennas for wideband applications, antenna and material properties measurements, and hardware and software acceleration of computational techniques for electromagentics. Dr. Elsherbeni is the co-author of *The Finite Difference Time Domain Method for Electromagnetics With MATLAB Simulations* (SciTech 2009), *Antenna Design and Visualization Using Matlab* (SciTech, 2006), *MATLAB Simulations for Radar Systems Design* (CRC Press, 2003), *Electromagnetic Scattering Using the Iterative Multiregion Technique* (Morgan & Claypool, 2007), *Electromagnetics and Antenna Optimization using Taguchi's Method* (Morgan & Claypool, 2007), and the main author of the chapters "Handheld Antennas" and "The Finite Difference Time Domain Technique for Microstrip Antennas" in *Handbook of Antennas in Wireless Communications* (CRC Press, 2001). Dr. Elsherbeni is a fellow member of the Institute of

Electrical and Electronics Engineers (IEEE) and a fellow member of the Applied Computational Electromagnetic Society (ACES). He is the Editor-in-Chief for the *ACES Journal*.

ERCUMENT ARVAS

Ercument Arvas received B.S. and M.S. degrees from the Middle East Technical University, Ankara, Turkey, in 1976 and 1979, respectively, and a Ph.D. from Syracuse University, Syracuse, NY, in 1983, all in electrical engineering. From 1984 to 1987, he was with the Electrical Engineering Department, Rochester Institute of Technology, Rochester, NY. In 1987, he joined the Electrical Engineering and Computer Science Department, Syracuse University, where he is currently a professor. His research and teaching interests are in computational electromagnetic radiation and scattering, and microwave devices. He is a member of the Applied Computational Electromagnetics Society (ACES) and a fellow of IEEE.

Printed in the United States
by Baker & Taylor Publisher Services